Recycling of Biomass Ashes

Heribert Insam • Brigitte A. Knapp
Editors

Recycling of Biomass Ashes

Springer

Editors
Prof.Dr. Heribert Insam
Universität Innsbruck
Inst. Mikrobiologie
Technikerstr. 25
6020 Innsbruck
Austria
heribert.insam@uibk.ac.at

Dr. Brigitte A. Knapp
Universität Innsbruck
Inst. Mikrobiologie
Technikerstr. 25
6020 Innsbruck
Austria
b.knapp@uibk.ac.at

ISBN 978-3-642-19353-8 e-ISBN 978-3-642-19354-5
DOI 10.1007/978-3-642-19354-5
Springer Heidelberg Dordrecht London New York

Library of Congress Control Number: 2011932977

Cover photo and artwork: Helene Keller

Cover design: deblik, Berlin

Printed on acid-free paper

Springer is part of Springer Science+Business Media (www.springer.com)

Preface

Meaningful Use of Biomass Ashes: Closing the Cycle

Biomass ash is the solid residue from the combustion of plant biomass which is increasingly being used for heat and electricity production. Biomass ash contains a wealth of macronutrients and micronutrients. Despite the value of the various elements contained in the ashes, their disposal in landfills is still common practice, generating considerable costs for biomass plant operators and negating the recycling potential of ashes. A prerequisite for sustainable use of ashes in agriculture and forestry, however, is their quality in terms of nutrients, on the one hand, and of heavy metals and organic pollutants, on the other. Appropriate combustion and separation techniques to obtain qualitatively valuable ash fractions are thus highly desirable.

To bring together knowledge and ideas on the reutilization of biomass ashes, the conference "Recycling of Biomass Ashes" was held in Innsbruck in March 2010, focusing on various recycling technologies for biomass ashes. This book comprises 11 chapters that are based on selected conference contributions. An introductory chapter by Insam and Knapp gives an overview of current technologies and future needs for ash recycling. In Chap. 2 (Schiemenz et al.), the virtue of ashes as a phosphorous source is emphasized. In Chaps. 3, 5 and 6, nutrient-related aspects of ashes from wood (Haraldsen et al. and Omil et al.) and olive residues (Nogales et al.) are addressed. Nieminen in Chap. 4 addresses the effect of wood ash on the soil fauna. Bougnom et al. in Chap. 7 and Sarabèr et al. in Chap. 8 discuss the potential of tropical acid soil melioration with ashes (from wood and cocoa residues, respectively). In Chaps. 9 and 10, Mödinger and Berra et al. elaborate on the potential of wood ash use in brick making and in the cement industry, respectively. Ash recycling as a puzzle stone in a sustainable society is addressed by Ribbing and Bjurström in Chap. 11.

The editors gratefully acknowledge the assistance of the many reviewers in improving the manuscripts and thank Springer-Verlag for excellent cooperation during the production of the book. The support of the Tiroler Zukunftsstiftung (K-Regio Center BioTreaT) is also acknowledged.

Innsbruck, Austria

Heribert Insam
Brigitte A. Knapp

Contents

Chapter 1
Recycling of Biomass Ashes: Current Technologies and Future Research Needs

Brigitte Amalia Knapp and Heribert Insam

Abstract Biomass ash is a final by-product from biomass incineration and is being produced in increasing amounts. Ash contains a variety of macronutrients and micronutrients and thus requires an appropriate recycling strategy. This chapter addresses various recycling strategies and technologies, with a particular focus on a smart combination of wastes from different sources for optimising recycling efficiency.

1.1 Introduction

Biomass ash is the solid residue accumulating from the thermal combustion of plant biomass for heat and electricity production, containing a variety of macronutrients and micronutrients resistant to incineration. As combustion of biomass is among the dominant bioenergy applications worldwide, increasing numbers of biomass-based power plants are being built and thus vast quantities of ashes are produced. Despite the value of the various elements contained in the ashes, their disposal in landfills is still common practice, generating considerable costs for biomass plant operators and negating the recycling potential of ashes. A prerequisite for sustainable use of ashes in agriculture and forestry, however, is their quality in terms of heavy metal contents and organic pollutants. Appropriate combustion and separation techniques for the different ash fractions are thus highly needed.

To bring together knowledge and ideas on the reutilisation of biomass ashes, a conference entitled "Recycling of Biomass Ashes" was held in Innsbruck in March 2010, focusing on various recycling technologies for biomass ashes. The conference sessions were targeted at the use of ashes as fertiliser or a supplement for organic and inorganic fertilisers as well as their combination with compost and anaerobic sludges. Further, ash amendments to forest soils were a major topic, as was the use of ashes for geotechnical constructions and industrial processes.

B.A. Knapp (✉) and H. Insam
University of Innsbruck, Institute of Microbiology, Technikerstraße 25, 6020 Innsbruck, Austria
e-mail: b.knapp@uibk.ac.at

H. Insam and B.A. Knapp (eds.), *Recycling of Biomass Ashes*,
DOI 10.1007/978-3-642-19354-5_1,

Moreover, national and international policies regulating the application of ash were presented and joint programmes for advancing knowledge in the field of ash recycling were discussed.

In the following chapters an overview of different fields of application for biomass ashes is presented, summarising current knowledge on the reutilisation of biomass ashes and highlighting future research needs. As most investigations on ash recycling are based on wood ash, this chapter will focus on ash produced from wood combustion.

1.2 Characterisation of Ash

During incineration of wood and other types of plant biomass, a solid residue is formed, representing about 2% (e.g. willow wood) to 20% (e.g. rice husks) of the input material (Jenkins et al. 1998). Depending on the plant species, the origin of plant, the plant parts used for combustion, the process parameters during incineration and the storage conditions of combustion residues, ashes differ considerably regarding their physical and chemical properties (Demeyer et al. 2001). These characteristics determine the quality of different ash types and their suitability for further applications (Karltun et al. 2008). Moreover, different treatments after combustion (self-hardening, thermal treatment or hardening with the addition of a binding material such as a potassium silicate) affect leaching properties of the ash. Ash pellets with a denser structure and a smaller specific surface area display lower leaching rates (Mahmoudkhani et al. 2007). The application form of biomass ash is of great concern, as untreated ash is difficult to apply evenly to soil and may lead to burning of the plant surfaces. Pretreatment of ash may thus be necessary to prevent such damage by lowering the reactivity of the ash. Pretreated ash products are assumed to be more suitable for application purposes, result in less dust formation during spreading, facilitate even spreading and prolong the fertiliser effect owing to slower decomposition rates (Sarenbo et al. 2009).

1.2.1 Wood Ash Composition

Although carbon (C) is mostly oxidised and nitrogen (N) is emitted in the form of gaseous compounds during combustion, most other elements present in the plant material are retained in the ash (Steenari et al. 1999). Wood ash mainly consists of calcium, potassium, magnesium, silicon, aluminium, phosphorus, sodium, manganese and sulphur, whereas it is N-deficient. Trace elements found in wood ash are iron, zinc, arsenic, nickel, chromium, lead, mercury, copper, boron, molybdenum, vanadium, barium, cadmium and silver, again found in varying concentrations in different types of wood ash (Demeyer et al. 2001; Karltun et al. 2008). The behaviour patterns of these elements differ considerably, as some elements are partially or completely volatilised during combustion, whereas others are retained to a high degree (Miller et al. 2002). Owing to the incomplete combustion of biomass, remaining C can be found in the ash to some extent, usually as charcoal (Karltun et al. 2008). Whereas the amounts of K, S, B, Na and Cu were reported to decrease with furnace temperature, the amounts of Mg, P, Mn, Al, Fe, Si and Ca were not affected by temperature (Misra et al. 1993). However, these effects depend on the tree species incinerated (Pitman 2006). Moreover, a lack of standardisation concerning the methods used for the assessment of major and minor ash-forming elements causes further divergences (Baernthaler et al. 2006).

1.2.2 Heavy Metals and Organic Pollutants in Wood Ash

The heavy metals that may accumulate in wood ash are of special concern when it is used for fertilisation purposes. Compared with coal ashes, ashes derived from wood are lower in heavy metals, but are more alkaline (Campbell 1990). High concentrations of As, Cd, Cr, Pb, Zn and Cu may, however, occur owing to the incineration of surface-treated waste wood and wood treated with industrial preservatives (Krook et al. 2006). Cu concentrations in biomass ashes were frequently shown to exceed critical values according to national regulations in Austria (Neurauter et al. 2004) and Germany (Ministerium für Umwelt und Verkehr Baden-Württemberg 2003). Average Cu contents found in three studies dealing with ash composition are illustrated in Fig. 1.1 (Neurauter et al. 2004; Niederberger 2002; Tóthóva 2005). Whereas the incineration of pure wood led to moderate Cu concentrations in the resulting ash, high Cu contents were found when other biomass, especially roadside greenery and material derived from wood processing, was combusted together with natural wood (samples 9, 10).

Wood ash is better applicable for fertilisation purposes if it is separated into fly and bottom ash during combustion, as heavy metals – except for Zn – accumulate in the fly ash (Pitman 2006; Stockinger et al. 2006). Fly ash is the lightest fraction formed during combustion, being deposited within the boiler or in the filters (Pitman 2006). Ashes may also include organic pollutants such as polychlorinated dibenzodioxin, biphenyls, dibenzofuran and polycyclic aromatic hydrocarbons (PAHs), which are of interest because of their toxic, mutagenic and carcinogenic effects (Lavric et al. 1994; Enell et al. 2008). High amounts of PAHs are ascribed to a poor combustion performance (Sarenbo 2009). Wood ashes may pose a risk not only because of the direct input of organic pollutants, but also because a rise in soil pH following wood ash amendments enhances remobilisation of PAHs and polychlorinated biphenyls (Bundt et al. 2001). An elevated pH also affects metal solubility in soil; however, changes in solubility do not necessarily correlate with incorporation of heavy metals in plants grown on the respective soils (Dimitriou et al. 2006). Another essential issue in regard of ash amendments to soils is leaching of toxic substances to the groundwater (Williams 1997), especially in combination with an elevated pH and high Na content (Morris et al. 2000). Leaching is frequently evaluated in laboratory tests, but

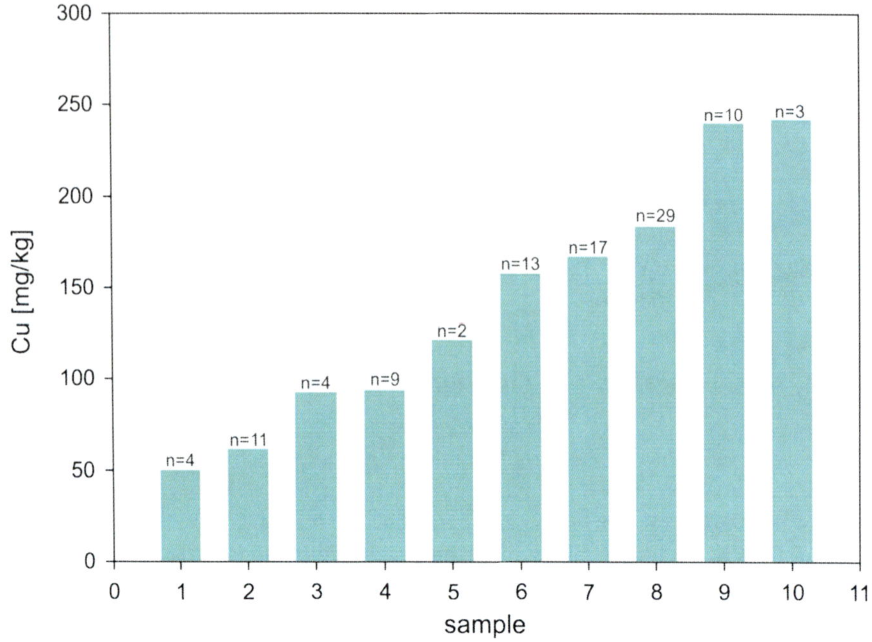

Fig. 1.1 Cu concentration (mg/kg) detected in a variety of ash samples derived from plain wood incineration, or combined incineration of wood with other biomass. (Data compiled from Neurauter et al. 2004, sample 4; Niederberger 2002, samples 5–10; Tóthóva 2005, samples 1–3)

these tend to overestimate or underestimate *on-site* leaching processes and thus it is difficult to assess the real situation in the field (Reijnders 2005).

1.3 Areas of Application for Wood Ashes

As versatile as wood ash is, its potential areas of application are:

- Ash application in forest ecosystems

  Wood ash is commonly applied to forest ecosystems to return nutrients extracted through whole-tree harvesting and to counteract soil acidification (Sect. 1.3.1).

- Wood ash as fertiliser or fertiliser supplement in agroecosystems

  Wood ash rich in nutrients but displaying a low concentration of heavy metals or organic pollutants is also suitable as fertiliser or fertiliser supplement for agricultural and horticultural purposes (Sect. 1.3.2).

- Wood ash for geotechnical constructions and industrial processes

Typical applications in this field are the construction of roads and parking areas, the use of ash as a surface layer in landfills and admixture of ash for concrete, brick or cement production (Sect. 1.3.3).

1.3.1 Ash Application in Forest Ecosystems

The effect of wood ash application on forest ecosystems has been intensively studied in northern European countries where ash is used as fertiliser in boreal forests (Aronsson and Ekelund 2004). Owing to extensive forest harvesting (especially whole-tree harvesting), reuse of ashes was established to avoid base element depletion of forest soils, leading to increasing acidity as well as decreasing amounts of nutrients and organic matter in the soil, thus threatening forest productivity (Stupak et al. 2008).

1.3.1.1 Effects of Wood Ash Application on Soil Properties

Wood ash is applied to forest soils to alleviate nutrient depletion and soil acidification, either alone or in combination with N fertiliser. Wood ash is also applied as lime replacement, providing base cations to increase soil pH (Steenari et al. 1999; Meiwes 1995; Brunner et al. 2004). This liming effect can be attributed to Ca and Mg carbonates in the ash as well as to its fine structure (Pitman 2006). Arvidsson and Lundkvist (2003) observed an increased soil pH after 3 Mg ha^{-1} wood ash application in young Norway spruce (*Picea abies*) stands. Moreover, concentrations of exchangeable Ca and Mg as well as the effective cation-exchange capacity were elevated compared with the control. As salts contained in the ash started to dissolve after application, high K, Na and SO_4 concentrations were also found in the soil (Augusto et al. 2008). Jacobson et al. (2004) reported an increased soil pH and base-cation content 5 years after amendment with self-hardened and crushed ash (3, 6 or 9 t ha^{-1}) or pelleted ash (3 t ha^{-1}) on two different coniferous sites in Sweden, whereby the ash formulation did not have an effect on soil chemistry despite differences in solubility. Basic substances used to amend soil may, however, foster nitrification and nitrate leaching in soil ecosystems and hence enhance soil acidity, counteracting the positive effects of wood ash application (Meiwes 1995). Since ash components bind to organic substances in the humus layer of forest soils, fertilisation effects of wood ash amendments on soil acidity and extractable Ca and Mg were found to last for many years (Bramryd and Fransman 1995; Saarsalmi et al. 2001, 2004, 2005; Mandre et al. 2006). The impact of wood ash applications (9 and 18 Mg ha^{-1}) on soil properties in different tree stands (European larch, aspen and poplar) was evaluated in a 7-year experiment in Michigan, revealing that wood ash was able to foster long-term productivity and repeated applications may even have the potential to make up for biomass-C losses due to plantation management operations (Sartori et al. 2007).

1.3.1.2 Effects of Ash Amendments on Trees and Ground Vegetation in Forest Ecosystems

When looking at the effect of wood ash amendments on tree growth in Nordic countries, Augusto et al. (2008) revealed a considerable site dependency using a meta-analysis approach. Whereas wood ash was not able to improve tree growth on mineral soils, it had a significant effect on trees planted on organic soils. Reviewing different studies from Finland, Sweden and Switzerland with regard to the impact of wood ash applications on tree growth and vitality, Lundström et al. (2003) reported neutral or even negative effects of ash fertilisation. Investigating the effect of hardened wood ash application (up to 3 Mg ha^{-1}) on ground vegetation in young Norway spruce stands on mineral soils, Arvidsson et al. (2002) found that biodiversity and plant biomass were not affected. In a Swiss forest, fine roots of spruce were influenced by ash application (4 t ha^{-1}) on mineral soil, whereby ash fertilisation enhanced the number of root tips, forks and the root length, but resulted in decreased root diameters (Genenger et al. 2003). In a set of field experiments applying wood ash (1–9 Mg ha^{-1}) on 30–60-year-old Scots pine (*Pinus sylvestris* L.) and Norway spruce (*Picea abies* (L.) Karst.) stands on mineral soil in Sweden, stem growth was only promoted when N (150 kg ha^{-1}) was added, whereas wood ash amendments without N did not lead to significant responses (Jacobson 2003). The same was true for combined wood ash and N applications on a Scots pine (*Pinus sylvestris*) stand on a low-productivity mineral soil, where wood ash plus N positively influenced tree growth even 20 years after application. Nitrogen fertilisation alone only led to a short-term effect (Saarsalmi et al. 2006). Whereas wood ash or sludge application alone did not have any influence on the structure of a commercial willow plantation in central Sweden in a 3-year experiment, harvestable shoot biomass was increased by a combined sludge (2.6 t ha^{-1}) and ash (5.5 t ha^{-1}) treatment and thus gave results comparable to fertilisation with mineral fertiliser corresponding to 14.5 kg P ha^{-1} $year^{-1}$, 48 kg K ha^{-1} $year^{-1}$ and 100 kg N ha^{-1} $year^{-1}$ (Adler et al. 2008). However, this treatment showed negative effects on wood fuel quality concerning P, K and heavy metal concentrations in the bark and wood. Plant growth or biomass production was also not influenced by wood ash fertilisation (10 and 20 Mg ha^{-1} for 3 years) in a willow plantation on a silt loam soil in the state of New York (Park et al. 2005). In this experiment, wood ash did not have an impact on nutrient concentrations of foliar, litter and stem tissue, whereas the concentrations of soil-extractable P, K, Ca and Mg were significantly higher than in control plots. In contrast, Moilanen et al. (2002) observed a positive effect of ash fertilisation (8 and 16 t ha^{-1}) on tree volume growth even 50 years after amendment on a drained peat mire, being accompanied by elevated nutrient concentrations in the peat.

The results presented are attributed to the fact that ash is low in N, which is the main limiting element for plant growth on mineral soils in boreal forests. In contrast, ash fertilisation was considered to be more suitable for peatland forests displaying higher N contents (Hånell and Magnusson 2005) and was found to promote tree growth (height, diameter and biomass) of a young *Pseudotsuga menziesii* plantation and a *Pinus radiata* plantation on N-rich mineral soil in

Spain (Solla-Gullón et al. 2006, 2008). Besides enhanced stem-wood growth of Norway spruce, Rosenberg et al. (2010) also observed increased CO_2 evolution rates even 12 years after wood ash applications (6 Mg ha^{-1}) on a N-rich soil, indicating that ash amendments of N-rich sites have to be evaluated carefully regarding their effect on C and N cycling.

1.3.1.3 Effects of Ash Amendments on Soil Microorganisms

Fertilisation of coniferous forest soil with wood ash (5 t ha^{-1}) was demonstrated to affect microbial biomass (on the basis of phospholipid fatty acid analysis). Fungi reacted more sensitively to wood ash treatment than bacteria, which was reflected by decreasing fungal-to-bacterial phospholipid fatty acid ratios (Baath et al. 1995). Because ectomycorrhizal fungi are known to play an important role in the nutrient supply of trees, Hagerberg and Wallander (2002) investigated the effect of wood ash amendment on a Norway spruce forest floor and revealed an increase in ectomycorrhizal biomass. The ectomycorrhizal fungus *Piloderma* sp. was found to frequently colonise granulated wood ash in a wood-ash-fertilised spruce forest, suggesting a direct impact on nutrient mobilisation (Mahmood et al. 2001, 2002). *Piloderma* sp. was moreover assumed to affect short-term storage of Ca derived from wood ash granules, whereas no effect on P storage or release was discovered (Hagerberg et al. 2005). Gaitnieks et al. (2005) reported a positive effect of wood ash (6 t ha^{-1}) on *Suillus* sp. when it was applied prior to planting of Scots pine seedlings; this was accompanied by increased root and needle biomass of the seedlings.

In a long-term study on different forest sites, wood ash fertilisation (5–8 t ha^{-1}) led to increased CO_2 production caused by enhanced microbial activity but did not influence N_2O emissions, although nitrification and denitrification may have been affected by wood ash application (Maljanen et al. 2006a, b). This effect was shown by Ozolincius et al. (2006) in a *Pinus sylvestris* stand in Lithuania, revealing an increase of ammonifying, nitrifying and denitrifying microorganisms after wood ash application (1.25–5 t ha^{-1}). In contrast, Saarsalmi et al. (2010) did not find changes in net nitrification when investigating the effect of wood ash (3 t ha^{-1}) combined with N (0.15 t ha^{-1}) on soil microbial processes in two coniferous stands in Finland 15 years after application.

1.3.1.4 Effects of Ash Amendments on Soil Fauna

Wood ash application at rates of 1 and 5 t ha^{-1} in a Scots pine stand in central Finland decreased numbers of the enchytraeid worm *Cognettia sphagnetorum*, and slightly changed the soil microarthropod community. Soil chemical parameters were also influenced by these treatments, whereas microbial communities were only affected by the higher ash concentration (Haimi et al. 2000). Enchytraeid size and abundance were found to be reduced in microcosms containing 30 g humus from a Norway spruce forest and amended with 480 mg wood ash, but the negative effect could be offset by sucrose, indicating that the impact of wood ash on soil animals in forest ecosystems is mainly linked to C input rates. Negative effects may thus be avoided by minimising C limitations for decomposers (Nieminen 2008; see Chap. 4, Nieminen 2011).

1.3.1.5 Contamination Risks Through Wood Ash Application in Forest Ecosystems

Heavy metal concentrations have to be considered when wood ash is recycled to forests; thus, the quality of the applied ash is of great concern to avoid accumulation of heavy metals in the environment (Stupak et al. 2008). In a microcosm experiment performed by Fritze et al. (2000), Cd derived from wood ash application on forest soils did not show any harmful effects on the microbial activity or fungal and bacterial community structure. In the same experiment, ash treatments were found to induce a shift in archaeal community patterns, whereas Cd alone or with ash did not have an influence (Yrjälä et al. 2004). When looking at the heavy metal contents of forest berries (*Rubus chamaemorus*, *Vaccinium vitis-idaea*, *Vaccinium uliginosum*) or mushrooms (*Russula paludosa*, *Lactarius rufus*, *Lactarius trivialis*, *Suillus variegates*, *Paxillus involutus*) on different Finish forest sites after wood ash fertilisation (4–14 t ha^{-1}), Moilanen et al. (2006) observed no accumulation of heavy metals or even a decrease in the long term. Similar results were found in six *Pinus radiata* plantations repeatedly fertilised with 4.5 t wood ash per hectare in northwestern Spain, leading to a decrease of Zn, Cu and Cd levels in some mushroom species, which was attributed to an increase in soil pH. Only Mn concentrations were elevated in all mushroom species investigated (*Amanita muscaria*, *Russula sardonia*, *Tricholoma pessundatum*, *Laccaria laccata*, *Micena pura*, *Suillus bovinus*, *Xerocomus badius*). Heavy metals did not accumulate in tree needles or ground vegetation (Omil et al. 2007). Another aspect of recycling ashes back to the soil is the accumulation of ^{137}Cs (Hedvall et al. 1996). However, application of wood ash at a level of 3.0 and 4.2 kg ha^{-1} contaminated with 30–4,800 Bq ^{137}Cs per kilogram on different coniferous forest sites in Sweden did not significantly increase radioactivity in the biological system (soil, field vegetation, tree parts) when measured 5–8 years after ash amendment, which was partly attributed to the antagonistic effects of wood ash K on ^{137}Cs (Högbom and Nohrstedt 2001). This finding was confirmed in a 100-year-old Scots pine (*Pinus sylvestris* L.) stand on Fe podsol in central Finland, on which ash fertilisation (1, 2.5 and 5 t ha^{-1}) led to a reduction of ^{137}Cs concentrations in lingonberries (*Vaccinium vitis-idaea* L.) analysed 2 and 7 years after application of ash (Levula et al. 2000).

1.3.2 Ash as Fertiliser or Fertiliser Supplement in Agroecosystems

Wood ash is not only a valuable fertiliser in forest ecosystems, it can also benefit agricultural soils, especially acid soil types.

Analysing the impact of wood ash (5 and 20 t ha^{-1}) on an Italian agricultural soil regarding its physicochemical, microbiological and biochemical properties, Perucci et al. (2008) observed increasing pH values and electrical conductivity as well as decreasing microbial biomass C in the first months after application, but no long-term effects of ash amendments were found. Enhanced crop production for barley (*Hordeum vulgare* L.) and canola oil seed (*Brassica rapa* L.) was monitored when Boralf soils in central Alberta were supplemented with wood ash (12.5 or 25 t ha^{-1}) in combination with N fertiliser (Patterson et al. 2004a). Although wood ash was, moreover, found to positively influence canola seed oil content, it may impair oil quality owing to an increase in the concentration of glucosinolate (Patterson et al. 2004b).

Combining wood ash with N sources is an interesting option for designer composts or fertilisers. Admixture of 8 and 16% wood ash and organic wastes prior to composting did have positive effects on the composting process (temperature, microbial activity) and the quality of the final product (no increase in heavy metal concentrations, improved nutrient balance) (Kuba et al. 2008). In comparison with mineral and organic fertilisers, wood-ash-amended compost was superior for the recultivation of a Tyrolean ski run, increasing plant cover and soil microbial biomass and respiration (Kuba et al. 2008) (Fig. 1.2). Composts produced with 8% wood ash admixture fostered utilisation of C sources (polymers, carboxylic and amino acids, alcohol, and carbohydrates) in a MicroResp™ assay and led to a change in microbial community structure, whereas compost with 16% ash altered bacterial and fungal community composition, but did not enhance C utilisation (Bougnom and Insam 2009). Bougnom et al. (2009, 2010) demonstrated that compost produced with wood ash supplement (8 and 16%) may be a cheap alternative to liming in tropical areas, where many soils are characterised by a low pH. Wood ash alone also significantly increased pH and electrical conductivity in a tropical soil in Cameroon and was found to supply nutrients to the soil (Voundi Nkana et al. 2002). Besides raising soil pH, wood ash amendments (4 and 6 t ha^{-1})

Fig. 1.2 Application of wood-ash-amended composts (8 and 16% w/w) in a reclamation trial on a ski slope in the Austrian Alps (Mutterer Alm, Tyrol, 1,700 m above sea level). The trial was set up in a randomised block with four replicates, including the two ash-amended composts as well as control plots and plots fertilised with organic or mineral fertilisers. (Photo: BioTreaT)

on an acid soil in Nigeria improved maize grain yield (Mbah et al. 2010). As lime or artificial fertilisers are unaffordable for resource-poor farmers, wood ash may be an alternative for improving soil fertility in agricultural soils in the tropics (Voundi Nkana et al. 1998; Bougnom et al. 2010; see Chap. 7, Bougnom et al. 2011). Wood ash application of up to 4 t ha^{-1} on tropical soil in Uganda was observed to enhance bean (*Phaseolus vulgaris*) and soybean (*Glycine max*. L) biomass, but led to higher Cu, Zn, Cd and Pb concentrations in edible plant parts (Mbaherekire et al. 2003). Analysing the effect of wood ash application as well as combined wood ash and compost amendments on soil microorganisms, Odlare and Pell (2009) revealed toxic effects of wood ash on potential denitrification in an arable soil on a short-term and a long-term basis. Compost was, however, able to mitigate these heavy-metal-related negative effects of the ash.

1.3.3 Recycling of Ashes for Geotechnical Constructions and Industrial Processes

Only high-quality ashes are suitable for the uses in agriculture and forestry explained in the previous sections. In particular for ashes that are characterised by elevated heavy metal contents, other uses are suggested, such as construction of roads, a surface layer in landfills, and as an additive in industrial processes such as concrete, brick, glass and cement production (van Alkemade et al. 1999; Ribbing 2007; Obernberger and Supanic 2009). Specific information on the use of biomass ashes in brick making is presented in Chap. 9 (Mödinger 2011), and in Chap. 10 Berra et al. (2011) focus on the reuse of woody biomass fly ash in cement-based materials. The use of ashes for civil works such as road construction or as a surface layer for landfills in Sweden is illustrated by Ribbing and Bjurström (see Chap. 11, Ribbing and Bjurström 2011).

1.4 Future Research Needs

The use of high-quality biomass ashes for fertilisation processes or for geotechnical and industrial purposes is advisable; however, there are several open questions for fostering different areas of application:

- If ash is to be used as fertiliser or fertiliser supplement on a large scale, it has to be turned into a marketable product in a cost-effective manner. Therefore, a standard product has to be developed, making it applicable on different sites with specific nutrient supply and pH regulation requirements (Table 1.1). Separation of high-quality and low-quality ashes at the incineration plant is highly desirable. Combustion processes that do not blend the entire ash but provide fractions of bottom and fly ash are preferable.

Table 1.1 Important steps for development and assessment of a user-friendly and cost-effective fertiliser with ash admixture

Product development and assessment
Formulation of the product
Detailed assessment of the processing conditions (granulation, pelletising)
Production of prototypes
Chemical assessment of prototypes
Analysis of effects of soil application regarding physicochemical and microbial parameters as well as plant productivity
Optimisation of product and procedures
Economic assessment and cost optimisation of product
Development of marketing strategies
Registration of product
Development of monitoring programmes (fertilisation effect, ecotoxicology)

- Ash applications to soils and the use of ash for geotechnical and industrial purposes have to be accompanied by quality controls to guarantee short-term and long-term harmlessness and the usefulness of these amendments/admixtures. International programmes have to be implemented to build up and merge knowledge on the recycling of biomass ashes and to work out general user guidelines, accompanied by standardised tests.
- Whereas wood ash application in forest ecosystems is commonly accepted in northern European countries, this is not so in other European countries. It is thus necessary to increase public awareness of the importance of sustainable forest management, including the recycling of wood ashes.
- On a national and European level, end-of-waste solutions for ashes should be envisaged, based on strict quality limits.

1.5 Conclusions

Biomass ash is the inorganic residue produced during incineration of biomass for heat and electricity production, containing valuable macronutrients and micronutrients from the combusted biomass. Physical and chemical characteristics of biomass ashes depend strongly on the plant type (plant species, origin of plants, plant parts combusted) as well as the process parameters during incineration and the storage conditions of ashes; thus, the quality of different biomass ashes differs considerably, even within the same incineration facility.

Investigations on the suitability of biomass ash application to forests as well as agroecosystems are difficult to compare, as different types and amounts of ash have been tested in various environments, hindering the establishment of general guidelines for ash amendments. Biomass ash applications have been shown to be beneficial to soil ecosystems, as long as ashes of good quality are used and care is taken not to accumulate heavy metals or organic pollutants as a consequence of high application doses or repeated applications. These positive effects have been confirmed in long-term experiments in forest ecosystems in northern European countries, where wood ash applications have a long tradition. Long-term investigations aside from these forest ecosystems are, however, scarce as are clear regulations and guidelines for biomass ash recycling to soil. The same is true for the use of combustion residues for geotechnical and industrial purposes.

Research on the recycling of biomass ashes is a relevant and timely issue. More specific information and recent research outcomes on the recycling of biomass ashes as fertiliser or fertiliser supplement in forests and agroecosystems as well as for geotechnical and industrial applications are provided in other chapters of this book.

Acknowledgements We wish to thank the University of Innsbruck, Austrian Federal Ministry of Science and Research, Federal Ministry of Agriculture, Forestry, Environment and Water Management, Land Tirol and Tiroler Zukunftsstiftung, Österreichische Bodenkundliche Gesellschaft, Tiroler Wasserkraft AG, Vorarlberger Kraftwerke AG, Salzburg AG and Dettendorfer Wertstoff GmbH & Co. KG as well as all project partners of K-Regio BioTreaT (http://www.biotreat.eu) for supporting the conference "Recycling of Biomass Ashes".

References

Adler A, Dimitriou I, Aronsson P, Verwijst T, Weih M (2008) Wood fuel quality of two *Salix viminalis* stands fertilised with sludge, ash and sludge–ash mixtures. Biomass Bioenergy 32:914–925

Aronsson KA, Ekelund NGA (2004) Biological effects of wood ash application to forest and aquatic ecosystems. J Environ Qual 33:1595–1605

Arvidsson H, Lundkvist H (2003) Effect of crushed wood ash on soil chemistry in young Norway spruce stands. For Ecol Manage 176:121–132

Arvidsson H, Vestin T, Lundkvist H (2002) Effect of crushed wood ash application on ground vegetation in a young Norway spruce stand. For Ecol Manage 161:75–87

Augusto L, Bakker MR, Meredieu C (2008) Wood ash applications to temperate forest ecosystems – potential benefits and drawbacks. Plant Soil 306:181–198

Baath E, Frostegrad A, Pennanen T, Fritze H (1995) Microbial community structure and pH response in relation to soil organic matter quality in wood-ash fertilized, clear-cut or burned coniferous forest soils. Soil Biol Biochem 27:229–240

Baernthaler G, Zischka M, Haraldsson C, Obernberger I (2006) Determination of major and minor ash-forming elements in solid biofuels. Biomass Bioenergy 30:983–997

Berra M, de Casa G, Dell'Orso M, Galeotti L, Mangialardi T, Paolini AE, Piga L (2011) Reuse of woody biomass fly ash in cement-based materials: leaching tests. In: Insam H, Knapp BA (eds) Recycling of biomass ashes. Springer, Heidelberg, pp 133–146

Bougnom BP, Insam H (2009) Ash additives to compost affect soil microbial communities and apple seedling growth. Bodenkultur 60:5–15

Bougnom BP, Mair J, Etoa FX, Insam H (2009) Composts with wood ash addition: a risk or a chance for ameliorating acid tropical soils? Geoderma 153:402–407

Bougnom BP, Knapp BA, Etoa FX, Elhottová D, Němcová A, Insam H (2010) Designer composts for ameliorating acid tropical soils: effects on the soil microbiota. Appl Soil Ecol 45:319–324

Bougnom BP, Knapp BA, Etoa FX, Insam H (2011) Possible use of wood ash and compost for improving acid tropical soils. In: Insam H, Knapp BA (eds) Recycling of biomass ashes. Springer, Heidelberg, pp 87–106

Bramryd T, Fransman B (1995) Silvicultural use of wood ashes – effects on the nutrient and heavy metal balance in a pine (*Pinus sylvestris*, L.) forest soil. Water Air Soil Pollut 85:1039–1044

Brunner I, Zimmermann S, Zingg A, Blaser P (2004) Wood-ash recycling affects forest soil and tree fine-root chemistry and reverses soil acidification. Plant Soil 267:61–71

Bundt M, Krauss M, Blaser P, Wilcke W (2001) Forest fertilization with wood ash: effect on the distribution and storage of polycyclic aromatic hydrocarbons (PAHs) and polychlorinated biphenyls (PCBs). J Environ Qual 30:1296–1304

Campbell AG (1990) Recycling and disposing of wood ash. TAPPI J 73:141–146

Demeyer A, Voundi Nkana JC, Verloo MG (2001) Characteristics of wood ash and influence on soil properties and nutrient uptake: an overview. Bioresour Technol 77:287–295

Dimitriou I, Eriksson J, Adler A, Aronsson P, Verwijst T (2006) Fate of heavy metals after application of sewage sludge and wood–ash mixtures to short-rotation willow coppice. Environ Pollut 142:160–169

Enell A, Fuhrman F, Lundin L, Warfvinge P, Thelin G (2008) Polycyclic aromatic hydrocarbons in ash: determination of total and leachable concentrations. Environ Pollut 152:285–292

Fritze H, Perkiomaki J, Saarela U, Katainen R, Tikka P, Yrjälä K, Karp M, Haimi J, Romantschuk M (2000) Effect of Cd-containing wood ash on the microflora of coniferous forest humus. FEMS Microbiol Ecol 32:43–51

Gaitnieks T, Brūvelis A, Indriksons A, Zālītis P (2005) Influence of wood ash on morphological parameters and mycorrhization of Scots pine containerized seedlings. For Stud 42:74–78

Genenger M, Zimmermann S, Hallenbarter D, Landolt W, Frossard E, Brunner I (2003) Fine root growth and element concentrations of Norway spruce as affected by wood ash and liquid fertilisation. Plant Soil 255:253–264

Hagerberg D, Wallander H (2002) The impact of forest residue removal and wood ash amendment on the growth of the ectomycorrhizal external mycelium. FEMS Microbiol Ecol 39:139–146

Hagerberg D, Pallon W, Wallander H (2005) The elemental content in the mycelium of the ectomycorrhizal fungus *Piloderma* sp. during the colonization of hardened wood ash. Mycorrhiza 15:387–392

Haimi J, Fritze H, Moilanen P (2000) Responses of soil decomposer animals to wood-ash fertilisation and burning in a coniferous forest stand. For Ecol Manage 129:53–61

Hånell B, Magnusson T (2005) An evaluation of land suitability for forest fertilization with biofuel ash on organic soils in Sweden. For Ecol Manage 209:43–55

Hedvall R, Erlandsson B, Mattsson S (1996) Cs-137 in fuels and ash products from biofuel power plants in Sweden. J Environ Radioact 31:103–117

Högbom L, Nohrstedt H-Ö (2001) The fate of ^{137}Cs in coniferous forests following the application of wood-ash. Sci Total Environ 280:133–141

Jacobson S (2003) Addition of stabilized wood ashes to Swedish coniferous stands on mineral soils – effects on stem growth and needle nutrient concentrations. Silva Fenn 37:437–450

Jacobson S, Högbom L, Ring E, Nohrstedt H-Ö (2004) Effects of wood ash dose and formulation on soil chemistry at two coniferous forest sites. Water Air Soil Pollut 158:113–125

Jenkins BM, Baxter LL, Miles TR Jr, Miles TR (1998) Combustion properties of biomass. Fuel Process Technol 54:17–46

Karltun E, Saarsalmi A, Ingerslev M, Mandre M, Andersson S, Gaitnieks T, Ozolinčius R, Varnagiryte-Kabasinskiene I (2008) Wood ash recycling – possibilities and risks. In: Röser D, Asikainen A, Raulund-Rasmussen K, Stupak I (eds) Sustainable use of forest biomass for energy. A synthesis with focus on the Baltic and Nordic region. Springer, Heidelberg, pp 79–108

Krook J, Mårtensson A, Eklund M (2006) Sources of heavy metal contamination in Swedish wood waste used for combustion. Waste Manage 26:158–166

Kuba T, Tschöll A, Partl C, Meyer K, Insam H (2008) Wood ash admixture to organic wastes improves compost and its performance. Agric Ecosyst Environ 127:43–49

Lavric ED, Konnov AA, De Ruyck J (2004) Dioxin levels in wood combustion – a review. Biomass Bioenergy 26:115–145

Levula T, Saarsalmi A, Rantavaara A (2000) Effects of ash fertilization and prescribed burning on macronutrient, heavy metal, sulphur and ^{137}Cs concentrations in lingonberries (*Vaccinium vitis-idaea*). For Ecol Manage 126:26–279

Lundström US, Bain DC, Taylor AFS, Van Hees PAW (2003) Effects of acidification and its mitigation with lime and wood ash on forest soil processes: a review. Water Air Soil Pollut Focus 3:5–28

Mahmood S, Finlay RD, Erland S, Wallander H (2001) Solubilisation and colonisation of wood ash by ectomycorrhizal fungi isolated from a wood ash fertilised spruce forest. FEMS Microbiol Ecol 35:151–161

Mahmood S, Finlay RD, Wallander H, Erland S (2002) Ectomycorrhizal colonisation of roots and ash granules in a spruce forest treated with granulated wood ash. For Ecol Manage 160:65–74

Mahmoudkhani M, Richards T, Theliander H (2007) Sustainable use of biofuel by recycling ash to forests: treatment of biofuel ash. Environ Sci Technol 41:4118–4123

Maljanen M, Jokinen H, Saari A, Strömmer R, Martikainen PJ (2006a) Methane and nitrous oxide fluxes, and carbon dioxide production in boreal forest soil fertilized with wood ash and nitrogen. Soil Use Manage 22:151–157

Maljanen M, Nykänen H, Moilanen M, Martikainen PJ (2006b) Greenhouse gas fluxes of coniferous forest floors as affected by wood ash addition. For Ecol Manage 237:143–149

Mandre M, Pärn H, Ots K (2006) Short-term effects of wood ash on the soil and the lignin concentration and growth of *Pinus sylvestris* L. For Ecol Manage 223:349–357

Mbah CN, Nwite JN, Njoku C, Nweke IA (2010) Response of maize (*Zea mays* L.) to different rates of wood-ash application in acid ultisol in Southeast Nigeria. Afr J Agric Res 5:580–583

Mbaherekire BJ, Oryem-Origa H, Kashambuzi J, Mutumba GM, Nyangababo JT (2003) Elemental composition of bean (*Phaseolus vulgaris*) and soy bean (*Glycine max* L.) grown on wood ash amended soil. Bull Environ Contam Toxicol 70:817–823

Meiwes KJ (1995) Application of lime and wood ash to decrease acidification of forest soils. Water Air Soil Pollut 85:143–152

Miller BB, Dugwell DR, Kandiyoti R (2002) Partitioning of trace elements during the combustion of coal and biomass in a suspension-firing reactor. Fuel 81:159–171

Ministerium für Umwelt und Verkehr Baden-Württemberg (2003) Schadstoffströme bei der Entsorgung von Holzasche. Reihe Abfall, Heft 76. http://www.uvm.baden-wuerttemberg.de/servlet/is/8774/Entsorgung%20von%20Holzasche.pdf?command=downloadContent&filename=Entsorgung%20von%20Holzasche.pdf. Accessed 2 Jun 2010

Misra MK, Ragland KW, Baker AJ (1993) Wood ash composition as a function of furnace temperature. Biomass Bioenergy 4:103–116

Mödinger F (2011) The use of biomass combustion ashes in brick making. In: Insam H, Knapp BA (eds) Recycling of biomass ashes. Springer, Heidelberg, pp 121–132

Moilanen M, Silfverberg K, Hokkanen TJ (2002) Effects of wood-ash on the tree growth, vegetation and substrate quality of a drained mire: a case study. For Ecol Manage 171:321–338

Moilanen M, Fritze H, Nieminen M, Piirainen S, Issakainen J, Piispanen J (2006) Does wood ash application increase heavy metal accumulation in forest berries and mushrooms? For Ecol Manage 226:153–160

Morris LA, Nutter WL, Miller WP, Overcash M (2000) Treatment and use of pulp and paper and textile industry residues in southern U.S. forests. In: Henry CL, Harrison RB, Bastian EK (eds) The forest alternative principles and practice of residuals reuse. University of Washington College of Forest Resources, Seattle, pp 208–217

Neurauter R, Mölgg M, Reinalter M (2004) Aschen aus Biomassefeuerungsanlagen. Amt der Tiroler Landesregierung Abteilung Umweltschutz/Referat Abfallwirtschaft. http://www.tirol.gv.at/fileadmin/www.tirol.gv.at/themen/umwelt/abfallwirtschaft/downloads/aschen-aus-biomassenfeuerungsanlagen-2004.pdf. Accessed 2 Jun 2010

Niederberger J (2002) Holzaschequalität und technische Rahmenbedingungen bei der Verbrennung – Ergebnisse einer Betreiberbefragung. In: Forstwirtschaftliche Fakultät der Universität Freiburg, Forstliche Versuchs- und Forschungsanstalt Baden-Württemberg (ed) Holzasche-Ausbringung im Wald, ein Kreislaufkonzept. Freiburger Forstliche Forschung, Heft 43. Eigenverlag FVA, Freiburg, pp 107–122

Nieminen JK (2008) Labile carbon alleviates wood ash effects on soil fauna. Soil Biol Biochem 40:2908–2910

Nieminen JK (2011) Wood ash effects on soil fauna and interactions with carbohydrate supply: a mini-review. In: Insam H, Knapp BA (eds) Recycling of biomass ashes. Springer, Heidelberg, pp 45–56

Obernberger I, Supanic K (2009) Possibilities of ash utilisation from biomass combustion plants. In: Proceedings of the 17th European Biomass Conference & Exhibition, June/July 2009, Hamburg, pp 2373–2384

Odlare M, Pell M (2009) Effect of wood fly ash and compost on nitrification and denitrification in agricultural soil. Appl Energy 86:74–80

Omil B, Piñeiro V, Merino A (2007) Trace elements in soils and plants in temperate forest plantations subjected to single and multiple applications of mixed wood ash. Sci Total Environ 381:157–168

Ozolincius K, Armolaitis K, Raguotis A, Varnagiryte I, Zenkovaite J (2006) Influence of wood ash recycling on chemical and biological condition of forest Arenosols. J For Sci 52:79–86

Park BB, Yanai RD, Sahm JM, Lee DK, Abrahamson LP (2005) Wood ash effects on plant and soil in a willow bioenergy plantation. Biomass Bioenergy 28:355–365

Patterson SJ, Acharya SN, Thomas JE, Bertschi AIB, Rothwell RL (2004a) Barley biomass and grain yield and canola seed yield response to land application of wood ash. Agron J 96:971–977

Patterson SJ, Acharya SN, Bertschi AIB, Thomas JE (2004b) Application of wood ash to acidic boralf soils and its effect on oilseed quality of canola. Agron J 96:1344–1348

Perucci P, Monaci E, Onofri A, Vischetti C, Casucci C (2008) Changes in physico-chemical and biochemical parameters of soil following addition of wood ash: a field experiment. Eur J Agron 28:155–161

Pitman RM (2006) Wood ash use in forestry – a review of the environmental impacts. Forestry 79:563–586

Reijnders L (2005) Disposal, uses and treatments of combustion ashes: a review. Resour Conserv Recycl 43:313–336

Ribbing CM (2007) Environmentally friendly use of non-coal ashes in Sweden. Waste Manage 27:1428–1435

Ribbing CM, Bjurström HG (2011) The Swedish ash programme with focus on bio-ashes. Ashes are a resource in a sustainable society. In: Insam H, Knapp BA (eds) Recycling of biomass ashes. Springer, Heidelberg, pp 147–164

Rosenberg O, Persson T, Högbom L, Jacobson S (2010) Effects of wood-ash application on potential carbon and nitrogen mineralisation at two forest sites with different tree species, climate and N status. For Ecol Manage 260:511–518

Saarsalmi A, Mälkönen E, Piirainen S (2001) Effects of wood ash fertilization on forest soil chemical properties. Silva Fenn 35:355–368

Saarsalmi A, Mälkönen E, Kukkola M (2004) Effect of wood ash fertilization on soil chemical properties and stand nutrient status and growth of some coniferous stands in Finland. Scand J Forest Res 19:217–233

Saarsalmi A, Derome J, Levula T (2005) Effect of wood ash fertilization on stand growth, soil water and needle chemistry, and berry yields of lingonberry (*Vaccinium vitis-idaea* L.) in a Scots pine stand in Finland. Metsanduslikud Uurim/For Stud 42:13–33

Saarsalmi A, Kukkola M, Moilanen M, Arola M (2006) Long-term effects of ash and N fertilization on stand growth, tree nutrient status and soil chemistry in a Scots pine stand. For Ecol Manage 235:116–128

Saarsalmi A, Smolander A, Kukkola M, Arola M (2010) Effect of wood ash and nitrogen fertilization on soil chemical properties, soil microbial processes, and stand growth in two coniferous stands in Finland. Plant Soil 331:329–340

Sarenbo S (2009) Wood ash dilemma-reduced quality due to poor combustion performance. Biomass Bioenergy 33:1212–1220

Sarenbo S, Mellbo P, Stålnacke O, Claesson T (2009) Reactivity and leaching of wood ash pellets dehydrated by hot air and flue gas. Open Waste Manage J 2:47–54

Sartori F, Lal R, Ebinger MH, Miller RO (2007) Tree species and wood ash affect soil in Michigan's Upper Peninsula. Plant Soil 298:125–144

Solla-Gullón F, Santalla M, Rodriguez-Soalleiro RJ, Merino A (2006) Nutritional status and growth of a young *Pseudotsuga menziesii* plantation in a temperate region after ash application of wood-bark ash. For Ecol Manage 237:312–321

Solla-Gullón F, Santalla M, Perez-Cruzado C, Merino A, Rodrıguez-Soalleiro RJ (2008) Response of *Pinus radiata* seedlings to application of mixed wood-bark ash at planting on temperate region: nutrition and growth. For Ecol Manage 255:3873–3884

Steenari B-M, Karlsson LG, Lindqvist O (1999) Evaluation of the leaching characteristics of wood ash and the influence of ash agglomeration. Biomass Bioenergy 16:119–136

Stockinger M, Kuba T, Plank R, Meyer K, Insam H (2006) Die Verwertung von Holzaschen-ein lösbares Problem? Österr Forstztg 117:20–21

Stupak I, Asikainen A, Röser D, Pasanen K (2008) Review of recommendations for forest energy harvesting and wood ash recycling. In: Röser D, Asikainen A, Raulund-Rasmussen K, Stupak I (eds) Sustainable use of forest biomass for energy. A synthesis with focus on the Baltic and Nordic region. Springer, Heidelberg, pp 155–196

Tóthóva S (2005) Effects of wood ash on soil solution and chemistry of leaves in a beech stand. In: Perchova J, Karas J (eds) Regular recycling of wood ash to prevent waste production. Proceedings of the RecAsh – international seminar, Prague, 8–9 Nov 2005, pp 35–46

van Alkemade MMC, Loo S, Sulilatu WF (1999) Exploratory investigation into the possibilities of processing ash produced in the combustion of reject wood. TNO-report TNO-MEP – R 99/357. TNO Institute of Environmental Sciences, Energy Research and Process Innovation. http://www.ieabcc.nl/publications/R99357.PDF. Accessed 28 Apr 2010

Voundi Nkana JC, Demeyer A, Verloo MG (1998) Chemical effects of wood ash on plant growth in tropical acid soils. Bioresour Technol 63:251–260

Voundi Nkana JC, Demeyer A, Verloo MG (2002) Effect of wood ash application on soil solution chemistry of tropical acid soils: incubation study. Bioresour Technol 85:323–325

Williams TM (1997) Obtaining water quality permits for land application of biomass broiler ash. Biomass Bioenergy 13:279–287

Yrjälä K, Katainen R, Jurgens G, Saarelaa U, Saanod A, Romantschuk M, Fritze H (2004) Wood ash fertilization alters the forest humus *Archaea* community. Soil Biol Biochem 36:199–201

Chapter 2
Phosphorus Fertilizing Effects of Biomass Ashes

Katja Schiemenz, Jürgen Kern, Hans-Marten Paulsen, Silvia Bachmann, and Bettina Eichler-Löbermann

Abstract The reutilization of biomass ashes in agriculture is important to create nutrient cycles. In field and pot experiments we investigated the fertilizing effects of different biomass ashes (rape meal ash, straw ash, and cereal ash) for eight different crops on a loamy sand and a sandy loam. Special emphasis was given to phosphorus (P). The ashes showed large differences in their elemental composition. The highest P contents (10.5%) were measured in the cereal ash, and lowest in straw ash (1% P). The solubility of P in water was low; however, about 80% of P was soluble in citric acid. Generally, the P fertilizing effect of ashes was comparable to that of highly soluble P fertilizers such as triple superphosphate. The ash supply resulted in an increase of P uptake of cultivated crops as well as in increased soil P pools (total P, water-soluble P, double-lactate-soluble P, oxalate-soluble P) and P saturation. The ash effects depended also on the cultivated crop. Good results were found in combination with phacelia, buckwheat, and maize. Provided that biomass ashes are low in heavy metals and other toxic substances, the ashes can be applied in agriculture as a valuable fertilizer.

2.1 Introduction

Renewable energy sources are important for reducing the EU's dependence on fossil fuels and cutting greenhouse gas emissions and other pollutants. "Biomass is one of the most important resources for reaching our renewable energy targets.

K. Schiemenz, S. Bachmann, and B. Eichler-Löbermann (✉)
Faculty of Agricultural and Environmental Sciences, University of Rostock, J.-von-Liebig-Weg 6, 18059 Rostock, Germany
e-mail: bettina.eichler@uni-rostock.de

J. Kern
Leibniz-Institut für Agrartechnik Potsdam-Bornim e.V., Max-Eyth-Allee 100, 14469 Potsdam, Germany

H.-M. Paulsen
Institute of Organic Farming vTI, Trenthorst 32, 23847 Trenthorst, Germany

H. Insam and B.A. Knapp (eds.), *Recycling of Biomass Ashes*,
DOI 10.1007/978-3-642-19354-5_2,

It already contributes more than half of renewable energy consumption in the European Union" said Günther Oettinger, Europe's energy commissioner in February 2010, when the European Commission adopted a report on sustainability requirements for the use of solid biomass and biogas in electricity, heating, and cooling (European Commission 2010).

In the context of increasing bioenergy production, the recycling of the residues in agriculture can contribute to realize nutrient cycles and reduce the necessity of commercial fertilizer application (see Chap. 3, Haraldsen et al. 2011; Chap. 7, Bougnom et al. 2011). This has special importance for phosphorus (P), since the P resources are strongly limited worldwide. Ashes from combustion of biomass are the oldest mineral fertilizer. Biomass ashes are nearly free of nitrogen but contain P and other nutrients needed for plant nutrition (Sander and Andrén 1997; Vance and Mitchell 2000; Patterson et al. 2004; Uckert 2004). Furthermore, biomass ashes can also be used as liming agents (Muse and Mitchell 1995; Mandre 2006), and can stimulate microbial activities in the soil (Demeyer et al. 2001; see Chap. 1, Knapp and Insam 2011).

The nutrient compositions of ashes are affected by different factors. Particularly, the kind of biomass combusted influences the quality and the nutrient values of ashes. Besides the raw material used, the combustion process itself affects the chemical composition of the ashes. The P concentration in biomass ashes may range from less than 1% up to 10% (Table 2.1).

Positive effects of biomass ashes on crop yields were found in different studies. Among others, Krejsl and Scanlon (1996) found wood ashes to increase oat and bean yields. Phongpan and Mosier (2003) found positive effects of rice hull ash on rice yield, Ikpe and Powell (2002) reported positive impacts of millet ash on millet yields, and Haraldsen et al. (2011, see Chap. 3) found particularly positive combination effects with meat and bone meal residues on barley.

Nutrient uptake efficiency and mobilization mechanisms of crops are important for high utilization of applied P (Schilling et al. 1998; Neumann 2007), and the utilization of P in ashes also depends on the cultivated crop. However, research findings concerning interactions between biomass ashes and crop species are rarely available.

In our studies we investigated three different crop biomass ashes and eight crop species within field and pot experiments. The objectives of our work were to evaluate the P fertilizing effect of biomass ashes on different soils, and to investigate possible interactions between the effects of ashes and cultivated crops.

Table 2.1 P contents in biomass ashes

Type of ash	P content (%)
Bagasse ash (Jamil et al. 2004)	0.01
Alfalfa stem ash (Mozaffari et al. 2002)	0.90
Horticulture ashes (Zhang et al. 2002)	0.04–1.00
Wood ashes (Erich and Ohno 1992; Saarsalmi et al. 2001; Hytönen 2003)	0.90–1.70
Wheat straw ash (Hytönen 2003)	1.30
Rape straw ash (Hytönen 2003)	2.10
Poultry litter ash (Yusiharni 2001; Codling et al. 2002)	5.00
Cereal ash (Eichler et al. 2008b)	10.4

2.2 Material and Methods

2.2.1 Treatments and Experimental Design

Two field experiments were conducted on different soil types at the agricultural experimental stations of the University of Rostock (loamy sand) and at the Institute of Organic Farming in Trenthorst (sandy loam). The same soil types were also used for the pot experiments in Rostock; in 2007 the experiments were established on loamy sand and in 2008 they were established on sandy loam (Table 2.2).

The rape meal ash (RMA) was produced at the University of Rostock in a fluidized bed combustion at 860°C. The rye straw ash (SA) was produced via grate firing at 750°C and was supplied by the Leibniz Institute for Agricultural Engineering in Potsdam-Bornim (Germany). The rye cereal ash (CA) was manufactured at the Agricultural Technical School of Tulln (Austria) also via grate firing at 650–850°C. The fertilization treatments in the field and pot experiments were established in respect of the nutrient contents of the ashes (Table 2.3). Heavy metal contents are given in Table 2.4.

Table 2.2 Soil characteristics at the beginning of the field and pot experiments

	Type of soil	pH	OM	Pw	Pdl	Pox	PSC	DPS
		$CaCl_2$	(%)	(mg kg^{-1})		(mmol kg^{-1})		(%)
Field experiments								
Rostock (0–30 cm)	Loamy sand	5.70	1.9	16.9	67.2	15.9	29.3	54.0
Trenthorst (0–30 cm)	Sandy loam	6.36	3.5	23.8	81.9	17.3	42.1	41.1
Pot experiments								
2007	Loamy sand	5.69	2.4	10.6	38.9	11.9	30.5	39.1
2008	Sandy loam	6.17	3.6	7.6	39.3	9.9	37.4	26.6

OM organic dry matter, *Pw* water-soluble P, *Pdl* double-lactate-soluble P, *Pox* oxalate-soluble P, *PSC* P sorption capacity, *DPS* degree of P saturation

Table 2.3 Treatments, nutrient concentrations of the ashes and nutrient supply, field and pot experiments

Fertilization treatments	Nutrient concentrations (%)			Field experiments				Pot experiments (6 kg soil per pot)			
				Fertilizer application rates (kg ha^{-1}) for 2 years	Nutrient supply (kg ha^{-1}) for 2 years			Fertilizer application rates (g pot^{-1})	Nutrient supply (g pot^{-1})		
	P	K	Mg		P	K	Mg		P	K	Mg
CON	–	–	–	–	–	–	–	–	–	–	–
Phosphorus (TSP)	20.2	–	–	–	–	–	–	1.00	0.20	–	–
RMA	8.0	7.3	5.5	650	51.7	47.5	35.4	2.50	0.20	0.18	0.14
SA	1.0	5.3	1.0	850	8.8	44.8	8.2	9.80	0.10	0.52	0.10
CA	10.5	10.8	3.3	500	52.4	54.2	16.7	1.90	0.20	0.21	0.06
Potassium (KCl)	–	52.4	–	–	–	–	–	1.00	–	0.52	–

CON control; *TSP* triple superphosphate; *RMA* rape meal ash; *SA* straw ash; *CA* cereal ash

Table 2.4 Heavy metal contents (mg kg^{-1}) and pH values of the biomass ashes

Biomass ash	pH	Cd	Cr	Cu	Hg	Ni	Pb	Zn
RMA	12.6	0.5	227.9	77.1	0.02	273.6	11.9	348.9
SA	11.1	0.1	4.7	24.5	0.02	3.7	<1.5	80.9
CA	12.9	1.3	13.7	170.9	0.04	13.1	2.6	750.5

RMA rape meal ash; *SA* straw ash; *CA* cereal ash

Table 2.5 Cultivated crops in the experiments in 2007 and 2008

Year	Field experiments		Pot experiments	
	Rostock	Trenthorst	Main crops	Catch crops
2007	Summer barley	Summer wheat	Maize, blue lupin, summer barley, oilseed rape	Oil radish, phacelia, Italian ryegrass, buckwheat
2008	Maize	Blue lupin		

In the 2-year field experiments with different crop plants (see Table 2.5) the three ashes were applied once at the beginning of the experiments and incorporated into the top soil. Nitrogen was given in all treatments according to good fertilization practice.

In the pot experiments six different fertilization treatments were established. Besides the ash treatments, other treatments included triple superphosphate (TSP) as a highly soluble P source, potassium chloride (KCl) as a highly soluble potassium source, and a control (CON) without P and potassium. The ashes/fertilizers were applied on the soil surface and mixed into the upper 5 cm of soil. For nitrogen supply, 1.4 g NH_4NO_3 per pot was given. Mitscherlich pots with 6 kg soil each were used for crop cultivation.

Eight different crops were investigated in the pot experiments. Depending on the favourable growing time of cultivated main crops and catch crops, two experiments per year were established (Table 2.5). The main crops were seeded in April and the catch crops were seeded in August. The crop growing period in the pot experiments was about 7–10 weeks until the time of maximum biomass. In field and pot experiments all treatments were replicated four times.

2.2.2 *Analyses*

Harvested shoots were dried at 60°C, weighed, ground with a plant mill, and stored for further investigations. The P content in plant tissue was measured after dry ashing using the molybdovanadate method (Page et al. 1982). Plant P uptake was calculated by multiplying the P content of the shoots and shoot biomass.

The soil samples were air-dried and sieved (2 mm) before analysis. Soil pH was measured in 0.01 M $CaCl_2$ using a 1:2.5 soil-to-solution ratio. For characterization of soil P pools, different methods were used. The method described by Van der Paauw (1971) was used to determine water-extractable P (Pw) with a soil-to-water ratio of 1:25. The P concentrations in the extracts were measured by the phosphomolybdate blue method via flow-injection analysis. The content of double-lactate-soluble P (Pdl) (photometric method) was quantified according to Blume et al. (2000). By means of the ammonium oxalate method (Schwertmann 1964) the extractable amount of P (Pox) allows the estimation of the amount of inorganic P being adsorbed on amorphous iron and aluminium oxides in the soil. Pox and the oxalate-soluble aluminium and iron contents (Alox, Feox) in soil were extracted and their concentrations were determined by inductively coupled plasma optical emission spectroscopy (ICP-OES; JY 238, Jobin Yvon, France). With use of these data, the P-sorption capacity [PSC (mmol kg^{-1}) = (Alox + Feox)/2] and the degree of P saturation [DPS (%) = Pox/PSC $\times$ 100] could be calculated according to Lookman et al. (1995) and Schoumans (2000). Total P was analysed after aqua regia dissolution in a microwave oven (Mars Xpress, CEM, Kamp-Lintfort, Germany) followed by ICP-OES.

Furthermore, the sequential P fractionation method developed by Hedley et al. (1982) was used. Different P fractions of decreasing bioavailability (resin-P, $NaHCO_3$-P, NaOH-P, H_2SO_4-P) were extracted step by step by using stronger extracting agents. The remaining P in the soil sample after the extraction steps is considered as residual P. The residual P content was determined by subtracting the amount of extracted P from the total P content [residual P = total P − (resin-P + $NaHCO_3$-P + NaOH-P + H_2SO_4-P)] as described by Schlichting et al. (2002). Total P was determined by aqua regia digestion in a microwave oven.

2.2.3 Statistics

Soil and plant data corresponding to four spatial replications were subjected to two- and one-factorial analysis of variance (general linear model). The results are reported as main effects and interactions. The means of soil and plant parameters were compared by the Duncan test. Significances were determined at $p \leq 0.05$. Significantly different means were indicated by using different characters. The statistical analysis was carried out using SPSS 15.0.

2.3 Results and Discussion

2.3.1 Effect of Biomass Ashes on P Uptake and Shoot Biomass

In the field experiments positive results of ash supply were found in Rostock on the loamy sand (Table 2.6), but not in Trenthorst on the sandy loam (Table 2.7). In Rostock in 2007 higher barley yields (significant) and higher P uptakes (by trend) were found after SA and RMA application in comparison with the control. For maize in 2008 the best effects were found again after SA supply but also after CA supply (Table 2.6).

Table 2.6 Effect of biomass ashes on yield and P uptake, Rostock field experiment (loamy sand)

Fertilization	Summer barley (grain) 2007		Maize (whole plant) 2008	
	Yield (FM, 14% water) (dt ha^{-1})	P uptake (kg ha^{-1})	Yield (DM) (dt ha^{-1})	P uptake (kg ha^{-1})
	0.039*	0.261 NS	0.012*	0.007**
CON	30.2 a	11.6	162 a	32.2 a
SA	35.3 b	13.0	180 b	39.2 c
RMA	35.5 b	13.1	165 a	33.3 ab
CA	33.6 ab	12.7	179 b	37.0 bc
Mean	33.6	12.6	172	35.4

Different characters indicate significant different means at $p \leq 0.05$ within a column
*$p \leq 0.05$; **$p \leq 0.01$
FM fresh matter, *DM* dry matter, *CON* control, *SA* straw ash, *RMA* rape meal ash, *CA* cereal ash, *NS* not significant

Table 2.7 Effect of biomass ashes on yield and P uptake, Trenthorst field experiment (sandy loam)

Fertilization	Summer wheat (grain) 2007		Blue lupin (grain) 2008	
	Yield (FM, 14 % water) (dt ha^{-1})	P uptake (kg ha^{-1})	Yield (FM, 14% water) (dt ha^{-1})	P uptake (kg ha^{-1})
	0.370 NS	0.418 NS	0.184 NS	0.134 NS
CON	32.6	11.6	37.2	11.0
SA	31.5	11.3	37.0	10.7
RMA	31.1	11.1	33.6	9.8
CA	31.5	11.3	36.9	10.5
Mean	31.7	11.3	36.2	10.5

FM fresh matter, *CON* control, *SA* straw ash, *RMA* rape meal ash, *CA* cereal ash, *NS* not significant at $p \leq 0.05$

The missing effects in the Trenthorst experiment concerning yield and P uptake (Table 2.7) were most probably related to the soil conditions, mainly to the higher pH of this soil (Tables 2.2, 2.10, 2.11). Therefore, the liming effect of biomass ashes did not result in a further advantage regarding the availability of nutrients, like we expected for sandy soils with lower pH. Furthermore, the soil P content in the Trenthorst soil was higher, which may have masked the P fertilizing effects of the ashes.

Owing to the lower soil volume, the fertilizing effects were higher in the pot experiments than in the field experiments. Significant effects were found for both soils in the 2007 and 2008 experiments.

The crop P uptake increased when P was supplied, independently of whether ash or TSP was added. In 2007, maize showed the highest P uptake of all main crops, with a mean value of 91.3 mg pot^{-1}. In comparison with the control, the maize P uptake rose owing to P supply. The highest increasing rates were found for CA (about 34%) and TSP (about 44%) (Table 2.8). These results are in coherence with the biomass yield of maize (data not shown).

In 2008 on sandy loam, barley, which generated the highest biomass, also had the highest P uptake (especially after RMA supply: 94.6 mg pot^{-1}) (Table 2.9).

Table 2.8 Effect of biomass ashes on P uptake of the main crops and the catch crops in the pot experiments in 2007 (loamy sand)

Fertilization	Maize	Blue lupin	Summer barley	Oilseed rape	Oil radish	Phacelia	Italian ryegrass	Buckwheat
	P uptake of the shoots (mg pot^{-1}) (relative values in parentheses)							
	0.000***	0.073 NS	0.000***	0.000***	0.006**	0.038*	0.000***	0.567 NS
CON	75.6 a	25.3	59.1 a	43.8 a	64.2 abc	99.8 a	61.1 b	103.8
	(100)	(100)	(100)	(100)	(100)	(100)	(100)	(100)
TSP	109.2 d	32.5	79.3 d	62.4 d	77.5 c	127.2 b	86.1 e	116.3
	(144)	(128)	(134)	(142)	(121)	(127)	(141)	(112)
RMA	99.9 bc	35.8	75.1 cd	60.9 cd	75.0 bc	136.1 b	89.5 e	121.6
	(132)	(142)	(127)	(139)	(117)	(136)	(146)	(117)
SA	91.4 b	27.2	70.1 bc	55.2 bc	54.6 a	135.9 b	70.2 c	111.0
	(121)	(108)	(119)	(126)	(85)	(136)	(115)	(107)
CA	101.0 cd	25.9	61.7 a	49.3 ab	77.0 c	129.5 b	78.6 d	127.7
	(134)	(102)	(104)	(113)	(120)	(130)	(129)	(123)
KCl	70.8 a	27.5	66.1 ab	43.5 a	62.6 ab	116.4 ab	51.8 a	113.1
	(94)	(109)	(112)	(99)	(98)	(117)	(85)	(109)
Mean	91.3	29.0	68.5	52.5	68.5	124.1	72.9	115.6

Different characters indicate significant different means at $p \leq 0.05$ within a column $^{*}p \leq 0.05$; $^{**}p \leq 0.01$; $^{***}p \leq 0.001$
CON control, *TSP* triple superphosphate, *RMA* rape meal ash, *SA* straw ash, *CA* cereal ash

Table 2.9 Effect of biomass ashes on P uptake of main crops and catch crops in the pot experiments in 2008 (sandy loam)

Fertilization	Maize	Blue lupin	Summer barley	Oilseed rape	Oil radish	Phacelia	Italian ryegrass	Buckwheat
	P uptake of the shoots (mg pot^{-1}) (relative values in parentheses)							
	0.000***	0.000***	0.000***	0.000***	0.000***	0.000***	0.000***	0.000***
CON	44.3 a	53.6 ab	67.1 a	47.2 a	57.2 a	110.9 a	51.5 ab	110.5 bc
	(100)	(100)	(100)	(100)	(100)	(100)	(100)	(100)
TSP	69.2 c	61.6 bc	92.7 b	78.4 b	94.1 c	144.4 c	80.4 c	137.1 d
	(156)	(115)	(138)	(166)	(165)	(130)	(156)	(124)
RMA	71.7 cd	64.9 c	94.6 b	73.2 b	89.3 c	150.3 c	73.5 c	120.8 cd
	(162)	(121)	(141)	(155)	(156)	(136)	(143)	(109)
SA	56.7 b	56.3 bc	70.6 a	67.9 b	71.4 b	142.0 c	55.9 b	78.2 a
	(128)	(105)	(105)	(144)	(125)	(128)	(109)	(71)
CA	77.0 d	74.0 d	88.6 b	73.0 b	86.7 c	150.0 c	80.1 c	127.5 cd
	(174)	(138)	(132)	(155)	(152)	(135)	(156)	(115)
KCl	45.4 a	46.4 a	62.7 a	51.9 a	64.6 ab	123.5 b	47.8 a	86.6 ab
	(102)	(87)	(93)	(110)	(113)	(111)	(93)	(78)
Mean	60.7	59.5	79.4	65.2	77.2	136.8	64.9	110.1

Different characters indicate significant different means at $p \leq 0.05$ within a column $^{***}p \leq 0.001$
CON control, *TSP* triple superphosphate, *RMA* rape meal ash, *SA* straw ash, *CA* cereal ash

Maize and lupin showed notable positive reactions on CA fertilization, with up to 74% more P uptake than in the control. For the catch crops, the highest P uptakes were found for phacelia and buckwheat in both pot experiments in 2007 and 2008.

Crop-specific P utilizations from the P sources were also relevant. In 2007, usually the highest effects were found for TSP and RMA, whereas the effects of RMA were a little smaller than those of TSP. The opposite was found for lupin and phacelia, with slightly better results due to RMA. In consequence of the lower P concentration, SA application usually resulted in a lower crop P uptake than the other ashes. For oil radish after SA supply, even lower values were found than in the control without P. However, the P uptake of phacelia in the SA treatment was 36% higher than in the control. The SA effect on the P uptake of phacelia was even comparable to the RMA effect (Table 2.8). Differences in P uptake were also found after CA application, with high values for maize and rather low values for barley and lupin.

These effects underline the crop-specific mechanisms (see also Fig. 2.1) which should be considered when planning ash application within the crop rotation. Plant-specific adaptation mechanisms may warrant a sufficient P supply also under conditions of P deficiency in soil. For example, rape may excrete organic acids in the root zone, which is an effective strategy to increase P uptake mainly on soil with higher pH (Hoffland 1992). Buckwheat has been shown to have different P uptake efficiencies depending on soil conditions (Zhu et al. 2002). According to Van Ray and Van Diest (1979), different plant species differ in their behaviour with respect to uptake of cations. Excessive accumulation of cations within the plant can result in net excretion of H^+, and in a lowering of pH values in soil, as was found in our experiments after cultivation of phacelia and buckwheat. Besides such kinds of chemical modifications in the rhizosphere, morphological root adaptations of plants may also help to supply plants with P (Fig. 2.1).

High P uptake of catch crops used as green manure provides a high potential for P supply of succeeding crops, when the decomposing catch crop releases P. According to our results, buckwheat and phacelia, which had high P uptake rates when fertilized with ashes, could be suitable in this sense. Furthermore, a combination of green manure crops and ash application can also provide the soil with organic material, which ashes do not contain.

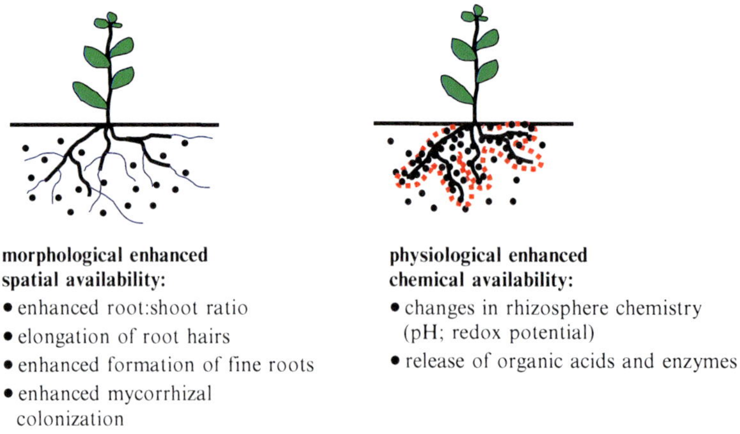

Fig. 2.1 Nutrient mobilization due to morphological and physiological mechanisms. (Modified from Neumann and Römheld 2002)

2.3.2 *Effect of Biomass Ashes and Cultivated Crops on Soil Characteristics*

In the field experiment in Rostock the bioavailable Pw and Pdl contents in soil (0–30 cm) were slightly increased following P supply (Table 2.10). Owing to a high standard deviation, these differences were not significant. The effects of the three ashes were similar. On average, slightly higher P values were measured for the CA treatment.

In Trenthorst the ashes also resulted in higher bioavailable P contents in soil (the differences were significant in spring 2008). Again, the highest increasing effect was found for the CA treatment (Table 2.11).

Like for the plant characteristics, the effects on the soil P pool were also found to be higher in the pot experiments than in the field experiments. Significant increases of high available P contents in the soil due to P application were measured at the end of the pot experiments.

In the pot experiments in 2007 with loamy sand the cultivated crops and interactions between crop and fertilization had also an effect. Cultivation of lupin resulted in the highest P values (Table 2.12). This can be partly explained by the

Table 2.10 Effect of biomass ashes on pH values and contents of water-soluble P and double-lactate-soluble P in the soil (0–30 cm), Rostock field experiment (loamy sand)

Fertilization	At harvest 2007			Spring 2008		At harvest 2008		
	pH	Pw (mg kg^{-1})	Pdl (mg kg^{-1})	Pw (mg kg^{-1})	Pdl (mg kg^{-1})	pH	Pw (mg kg^{-1})	Pdl (mg kg^{-1})
	0.398 NS	0.733 NS	0.701 NS	0.850 NS	0.692 NS	0.599 NS	0.840 NS	0.782 NS
CON	5.60	13.5	61.7	13.6	61.8	6.00	8.8	57.6
SA	5.60	15.3	64.3	14.7	63.7	5.96	9.4	59.8
RMA	5.63	13.9	64.3	14.1	62.9	5.96	9.7	60.0
CA	5.67	14.3	66.4	14.9	65.2	5.98	9.7	61.2
Mean	5.62	14.2	64.2	14.3	63.4	5.98	9.4	59.6

CON control, *SA* straw ash, *RMA* rape meal ash, *CA* cereal ash, *Pw* water-soluble P, *Pdl* double-lactate-soluble P, *NS* not significant at $p \leq 0.05$

Table 2.11 Effect of biomass ashes on pH values and contents of water-soluble P and double-lactate-soluble P in the soil (0–30 cm), Trenthorst field experiment (sandy loam)

Fertilization	At harvest 2007			Spring 2008		At harvest 2008		
	pH	Pw (mg kg^{-1})	Pdl (mg kg^{-1})	Pw (mg kg^{-1})	Pdl (mg kg^{-1})	pH	Pw (mg kg^{-1})	Pdl (mg kg^{-1})
	0.876 NS	0.689 NS	0.295 NS	0.590 NS	0.038*	0.909 NS	0.835 NS	0.236 NS
CON	6.39	21.9	80.7	18.2	80.0 a	6.46	18.6	76.3
SA	6.40	20.9	81.0	19.0	79.1 a	6.47	18.6	75.7
RMA	6.39	22.2	83.7	19.1	80.3 a	6.48	20.0	77.6
CA	6.44	22.7	84.6	20.3	84.6 b	6.49	19.6	80.1
Mean	6.41	21.9	83.8	19.2	81.0	6.47	19.2	77.4

Different characters indicate significant different means at $p \leq 0.05$ within a column $^{*}p \leq 0.05$
CON control, *SA* straw ash, *RMA* rape meal ash, *CA* cereal ash, *Pw* water-soluble P, *Pdl* double-lactate-soluble P, *NS* not significant

Table 2.12 Effect of biomass ashes and crops on pH values and contents of water-soluble P and double-lactate-soluble P in the soil (0–30 cm), pot experiments in 2007 (loamy sand)

Fertilizer	Maize	Blue lupin	Summer barley	Oilseed rape	Oil radish	Phacelia	Italian ryegrass	
	Buckwheat		pH					
	0.005**	0.036*	0.408 NS	0.336 NS	<0.001***	0.240 NS	0.002**	0.006**
CON	5.49 ab	5.54 ab	5.67	5.71	5.36 b	5.04	5.57 a	5.27 a
TSP	5.43 a	5.51 a	5.69	5.65	5.36 b	5.04	5.59 a	5.19 a
RMA	5.54 ab	5.53 a	5.69	5.67	5.47 bc	5.10	5.71 bc	5.37 a
SA	5.70 c	5.67 b	5.77	5.72	5.63 c	5.06	5.80 c	5.61 b
CA	5.58 bc	5.68 b	5.69	5.77	5.43 b	4.94	5.65 ab	5.33 a
KCl	5.44 ab	5.51 a	5.73	5.65	5.15 a	4.94	5.57 a	5.20 a
Mean	5.53	5.57	5.71	5.70	5.40	5.02	5.65	5.33
	Pw (mg kg^{-1})							
	0.000***	0.018*	0.000***	0.000***	0.000***	0.000***	0.025*	0.000***
CON	7.5 a	10.4 ab	7.8 a	8.0 a	8.7 a	8.1 a	9.6 a	8.0 a
TSP	10.7 bc	15.3 c	11.5 c	11.9 b	12.8 b	11.3 b	12.3 b	15.5 b
RMA	10.8 bc	12.7 abc	10.0 b	12.5 b	12.5 b	11.0 b	10.9 ab	15.7 b
SA	9.5 b	13.5 bc	10.1 b	11.5 b	15.5 c	10.9 b	10.6 ab	14.4 b
CA	11.9 c	11.7 abc	7.5 a	11.1 b	12.7 b	12.3 b	12.2 b	13.9 b
KCl	7.5 a	9.2 a	7.9 a	8.1 a	8.4 a	7.7 a	9.3 a	7.6 a
Mean	9.6	12.1	9.1	10.5	11.8	10.2	10.8	12.5
	Pdl (mg kg^{-1})							
	0.000***	0.000***	0.000***	0.000***	0.000***	0.000***	0.000***	0.000***
CON	33.0 a	34.7 a	33.3 a	33.8 a	31.7 a	26.4 a	33.7 a	29.5 a
TSP	39.5 b	44.9 b	41.9 b	41.2 b	39.2 b	32.1 bc	40.8 b	43.8 b
RMA	39.3 b	44.7 b	41.2 b	43.5 b	38.6 b	32.1 bc	42.0 b	42.5 b
SA	40.1 b	45.8 b	41.5 b	42.0 b	44.7 c	34.4 cd	42.0 b	45.2 b
CA	42.2 b	48.3 b	34.6 a	43.5 b	40.7 bc	35.5 d	41.2 b	41.5 b
KCl	32.8 a	34.9 a	34.1 a	32.2 a	30.6 a	30.4 b	34.2 a	31.3 a
Mean	37.8	42.2	37.8	39.4	37.6	31.8	39.0	39.0

Different characters indicate significant different means at $p \leq 0.05$ within a column $^{*}p \leq 0.05$; $^{**}p \leq 0.01$; $^{***}p \leq 0.001$

CON control, *TSP* triple superphosphate, *RMA* rape meal ash, *SA* straw ash, *CA* cereal ash, *Pw* water-soluble P, *Pdl* double-lactate-soluble P, *NS* not significant

low P uptake of this species (Table 2.8). Furthermore, in many studies P mobilization effects were found for lupin, albeit mainly for white lupin (Shen et al. 2003; Kania 2005) owing to special root morphology (cluster roots) and root-induced chemical changes in the rhizosphere. According to results of Egle et al. (2003) and Pearse et al. (2007), blue lupin, which was used in the pot experiments, does not generate such "cluster roots", but can enhance nutrient availability by means of carboxylate excretion into the soil and uptake of cations.

The decrease of the readily available P directly after phacelia harvest is probably only a temporary process. In a long-time field experiment with various catch crops, phacelia cultivation resulted in high levels of bioavailable P in the soil (Eichler-Löbermann et al. 2008a).

In the pot experiments with sandy loam (2008) the Pw and Pdl contents in the soil were also influenced by fertilization. The highest values were found after RMA, CA, and TSP application (Table 2.13).

Remarkably, in the experiments in 2007 and 2008 there were no differences in Pw contents of the soil between the P fertilizing treatments (TSP and ashes), even though commercial TSP fertilizer contains 80–93% water-soluble P (Mullins and Sikora 1995) and the water solubility of P in crop ashes is usually lower than 1% (Eichler-Löbermann et al. 2008b).

The P contents in soil were also related to the crop P uptake, namely high P uptakes usually resulted in P exhaustion in soil and in lower soil P contents. Thus, the relatively low P uptake of oil radish (Table 2.9) was related to rather high Pw and Pdl values (Table 2.13). Phacelia, which had the highest P uptake of all crops, showed comparably lower Pw and Pdl values. In contrast, high soil P contents were

Table 2.13 Effect of biomass ashes and crops on pH values and contents of Pw and Pdl in the soil (0–30 cm), pot experiments in 2008 (sandy loam)

Fertilizer	Maize	Blue lupin	Summer barley	Oilseed rape	Oil radish	Phacelia	Italian ryegrass	Buckwheat
	pH							
	<0.001***	<0.001***	0.006**	<0.001***	<0.001***	<0.001***	<0.001***	0.002**
CON	6.05 b	5.84 a	6.08 a	6.02 ab	5.79 a	5.51 a	6.09 a	5.81 a
TSP	5.98 a	5.79 a	6.06 a	5.98 a	5.83 ab	5.60 ab	6.15 ab	5.75 a
RMA	6.12 c	5.84 a	6.15 ab	6.07 b	5.95 b	5.70 b	6.27 cd	5.77 a
SA	6.20 d	6.02 b	6.26 c	6.18 c	6.15 c	5.83 c	6.41 e	6.03 b
CA	6.09 bc	5.84 a	6.21 bc	6.03 ab	5.93 b	5.63 b	6.23 bc	5.89 ab
KCl	6.09 bc	5.82 a	6.14 ab	6.06 b	5.95 b	5.51 a	6.32 d	6.00 b
Mean	6.09	5.86	6.15	6.06	5.93	5.63	6.24	5.88
	Pw (mg kg^{-1})							
	0.000***	0.000***	0.000***	0.000***	0.000***	0.000***	0.000***	0.000***
CON	6.1 a	6.5 a	5.9 a	7.3 a	6.6 a	6.5 a	5.7 a	5.7 a
TSP	9.8 bc	11.1 c	10.0 b	9.8 b	11.3 b	10.2 d	10.5 b	11.4 c
RMA	10.4 bc	11.2 c	9.5 b	10.4 b	12.2 b	9.1 c	11.0 b	11.2 c
SA	9.1 b	9.2 b	8.5 b	10.3 b	10.7 b	8.1 b	9.7 b	8.6 b
CA	11.0 d	11.3 c	9.0 b	9.7 b	11.4 b	10.2 d	10.0 b	10.7 bc
KCl	5.9 a	6.8 a	5.8 a	6.5 a	5.5 a	6.0 a	6.3 a	6.3 a
Mean	8.7	9.3	8.1	9.0	9.6	8.3	8.9	9.0
	Pdl (mg kg^{-1})							
	0.000***	0.000***	0.000***	0.000***	0.000***	0.000***	0.000***	0.000***
CON	34.4 a	34.5 a	33.0 a	34.1 a	28.2 a	26.8 a	32.1 a	28.3 a
TSP	44.1 b	46.0 cd	45.4 d	43.3 b	42.5 b	39.2 c	44.1 b	43.9 b
RMA	44.7 b	43.7 bc	42.4 bc	43.4 b	42.6 b	35.8 b	45.4 b	42.2 b
SA	43.8 b	41.3 b	41.6 b	44.3 b	41.3 b	35.8 b	45.8 b	43.3 b
CA	45.8 b	47.1 d	44.6 cd	44.7 b	46.9 c	39.7 c	44.2 b	45.6 b
KCl	34.9 a	33.2 a	34.3 a	33.2 a	29.3 a	27.8 a	33.8 a	32.8 a
Mean	41.3	41.0	40.2	40.5	38.5	34.2	40.9	39.4

Different characters indicate significant different means at $p \leq 0.05$ within a column
$^{**}p \leq 0.01$; $^{***}p \leq 0.001$
CON control, *TSP* triple superphosphate, *RMA* rape meal ash, *SA* straw ash, *CA* cereal ash, *Pw* water-soluble P, *Pdl* double-lactate-soluble P

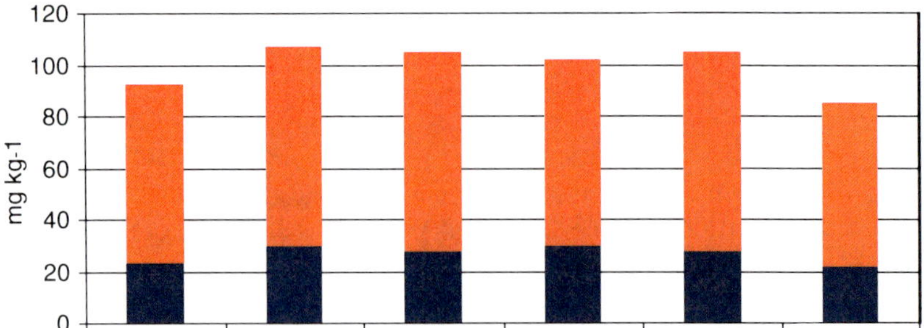

	CON	TSP	RMA	SA	CA	KCl
Pt	504	527	532	516	516	503
NaHCO3-P	69	77	77	72	77	64
Resin-P	23	30	27	30	28	22
sum % of Pt	18.3	20.3	19.7	19.7	20.3	17.0

Fig. 2.2 Average values of resin P and $NaHCO_3$ P in a pot experiment with main crops (maize, blue lupin, summer barley, oilseed rape) and loamy sand in 2007. *Pt* total P, *CON* control, *TSP* triple superphosphate, *RMA* rape meal ash, *SA* straw ash, *CA* cereal ash

found after buckwheat cultivation in the ash and TSP treatments (Table 2.13), despite the high P uptakes (Table 2.9).

The P fractionation method can provide additional information about the pathway of remaining ash P in the soil and helps to predict its presumable availability. In the pot experiments the highest soluble P fractions (resin P and $NaHCO_3$P) were increased by the ashes as well as by TSP, as was shown, for example, for the loamy sand (see Fig. 2.2). The less available P fractions (NaOH P, H_2SO_4 P, and residual P) however, were not affected (data not shown). In comparison with the control, the total P content increased when P was supplied (Fig. 2.2).

2.4 Conclusions

According to our results, a high fertilization effect of biomass ashes can be expected. In the pot experiments and in the field experiment in Rostock, biomass ashes led to raised P uptakes of the crops and increased contents of the more easily available P pools in soil (Pw, Pdl, resin P, $NaHCO_3$ P). Interactions of biomass ashes and cultivated crops had an additional effect on the utilization of P in ashes and should be considered for practical fertilization decisions. Provided that the ashes do not contain harmful substances, the utilization of biomass ashes in crop production is an important method for the recirculation of nutrients in agriculture and may save nutrient resources.

Acknowledgements This project was supported by the Federal Ministry of Food, Agriculture and Consumer Protection (BMELV), Germany (support code 22016206) (2007–2009). The project execution organization was the Agency for Renewable Resources (FNR), Germany. The project was accomplished in cooperation with the following project partners: the Institute of Organic Farming of the Johann Heinrich von Thünen-Institute (vTI)/Federal Research Institute for Rural Areas, Forestry and Fisheries, and the Leibniz Institute for Agricultural Engineering Potsdam-Bornim, Germany.

References

Blume HP, Deller B, Leschber R, Paetz A, Schmidt S, Wilke B-M (2000) Handbuch der Bodenuntersuchung. Terminologie, Verfahrensvorschriften und Datenblätter. Physikalische, chemische und biologische Untersuchungsverfahren. Gesetzliche Regelwerke. Wiley-VCH, Weinheim

Bougnom B, Knapp BA, Etoa FX, Insam H (2011) Possible use of wood ash and compost for improving acid tropical soils. In: Insam H, Knapp BA (eds) Recycling of biomass ashes. Springer, Heidelberg, pp 87–106

Codling EE, Chaney RL, Scherwell J (2002) Poultry litter ash as a potential phosphorus source for agricultural crops. J Environ Qual 31:954–961

Demeyer A, Nkana JCV, Verloo MG (2001) Characteristics of wood ash and influence on soil properties and nutrient uptake: an overview. Bioresour Technol 77:287–295

Egle K, Römer W, Keller H (2003) Exudation of low molecular weight organic acids by *Lupinus albus* L., *Lupinus angustifolius* L. and *Lupinus luteus* L. as affected by phosphorus supply. Agronomie 23:511–518

Eichler-Löbermann B, Köhne S, Kowalski B, Schnug E (2008a) Effect of catch cropping on phosphorus bioavailability in comparison to organic and inorganic fertilization. J Plant Nutr 31: 659–676

Eichler-Löbermann B, Schiemenz K, Makadi M, Vago I, Köppen D (2008b) Nutrient cycling by using residues of bioenergy production – II. Effects of biomass ashes on plant and soil parameters. Cereal Res Commun 36:1259–1262

Erich MS, Ohno T (1992) Phosphorus availability to corn from wood ash amended soils. Water Air Soil Pollut 64:475–485

European Commission (2010) Commission adopts biomass sustainability report. http://europa.eu/rapid/pressReleasesAction.do?reference=IP/10/192. Accessed 25 Feb 2010

Haraldsen TK, Pedersen PA, Grønlund A (2011) Mixtures of bottom wood ash and meat and bone meal as NPK fertilizer. In: Insam H, Knapp BA (eds) Recycling of biomass ashes. Springer, Heidelberg, pp 33–44

Hedley J, Stewart J, Chauhan B (1982) Changes in inorganic and organic soil phosphorus fractions by cultivation practices and by laboratory incubations. Soil Sci Soc Am J 46:970–976

Hoffland E (1992) Quantitative evaluation of the role of organic acid exudation in the mobilization of rock phosphate by rape. Plant Soil 140:279–289

Hytönen J (2003) Effects of wood, peat and coal ash fertilization on Scots Pine foliar nutrient concentrations and growth on afforested former agricultural peat soils. Silva Fenn 37(2):219–234

Ikpe FN, Powell JM (2002) Nutrient cycling practices and changes in soil properties in the crop-livestock farming systems of western Niger Republic of West Africa. Nutr Cycl Agroecosyst 62:37–45

Jamil M, Qasim M, Umar M, Subhan A (2004) Impact of organic wastes (bagasse ash) on the yield of wheat (*Triticum aestivum* L.) in a calcareous soil. Int J Agric Biol 6(3):468–470

Kania A (2005) Regulation of phosphate deficiency-induced carboxylate exudation in cluster roots of white lupin (*Lupinus albus* L.). Dissertation, University of Hohenheim

Knapp BA, Insam H (2011) Recycling of biomass ashes – current technologies and future research needs. In: Insam H, Knapp BA (eds) Recycling of biomass ashes. Springer, Heidelberg, pp 1–16

Krejsl JA, Scanlon TM (1996) Evaluation of beneficial use of wood-fired boiler ash on oat and bean growth. J Environ Qual 25:950–954

Lookman R, Vandeweert N, Merckx R, Vlassak K (1995) Geostatistical assessment of the regional distribution of phosphate sorption capacity parameters (Fe_{ox} and Al_{ox}) in northern Belgium. Geoderma 66:285–296

Mandre M (2006) Influence of wood ash on soil chemical composition and biochemical parameters of young Scots pine. Proc Estonian Acad Sci Biol Ecol 55(2):91–107

Mozaffari M, Russelle MP, Rosen CJ, Nater EA (2002) Nutrient supply and neutralizing value of alfalfa stem gasification ash. Soil Sci Soc Am J 66:171–178

Mullins GL, Sikora FJ (1995) Effect of soil pH on the requirement for water-soluble phosphorus in triple superphosphate fertilizers. Fertil Res 40:207–214

Muse JK, Mitchell CC (1995) Paper mill boiler ash and lime by-products as soil liming materials. Agron J 87:432–438

Neumann G (2007) Root exudates and nutrient cycling. In: Marschner P, Rengel Z (eds) Nutrient cycling in terrestrial ecosystems. Springer, Heidelberg, pp 123–157

Neumann G, Römheld V (2002) Root-induced changes in the availability of nutrients in the rhizosphere. In: Waisel Y, Eshel A, Kafkafi U (eds) Plant roots – the hidden half, 3rd edn. Dekker, New York, pp 617–649

Page AL, Miller RH, Keeney DR (1982) Methods of soil analysis. Agron no 9, part 2: chemical and microbiological properties, 2nd edn. American Society for Agronomy, Madison

Patterson SJ, Acharya SN, Thomas JE, Bertschi AB, Rothwell RL (2004) Integrated soil and crop management: barley biomass and grain yield and canola seed yield response to land application of wood ash. Agron J 96(4):971–977

Pearse SJ, Veneklaas EJ, Cawthray G, Bolland MD, Lambers H (2007) Carboxylate composition of root exudates does not relate consistently to a crop species' ability to use phosphorus from aluminium, iron or calcium phosphate sources. New Phytol 173(1):181–190

Phongpan S, Mosier AR (2003) Impact of organic residue management in nitrogen use efficiency in an annual rice cropping sequence of lowland central Thailand. Nutr Cycl Agroecosyst 66: 233–240

Saarsalmi A, Mälkönen E, Piirainen S (2001) Effects of wood ash fertilization on forest soil chemical properties. Silva Fenn 35(3):355–368

Sander M-L, Andrén O (1997) Ash from cereal and rape straw used for heat production: liming effect and contents of plant nutrients and heavy metals. Water Air Soil Pollut 93:93–108

Schilling G, Gransee A, Deubel A, Lezovic G, Ruppel S (1998) Phosphorus availability, root exudates, and microbial activity in the rhizosphere. Z Pflanzenernähr Bodenkd 161:465–478

Schlichting A, Leinweber P, Meissner R, Altermann M (2002) Sequentially extracted phosphorus fractions in peat-derived soils. J Plant Nutr Soil Sci 162:290–298

Schoumans OF (2000) Determination of the degree of phosphate saturation in non-calcarous soils. In: Pierzynski GM (ed) Methods for phosphorus analysis for soils, sediments, residuals, and water. Southern Cooperative Series Bulletin 396, North Carolina, pp 31–34

Schwertmann U (1964) Differenzierung der Eisenoxide des Bodens durch Extraktion mit Ammoniumoxalatlösung. Z Pflanzenernähr Bodenkd 105:194–202

Shen J, Rengel Z, Tang C, Zhang F (2003) Role of phosphorus nutrition in development of cluster roots and release of carboxylates in soil-grown *Lupinus albus*. Plant Soil 248:199–206

Uckert GB (2004) Versuche zur landbaulichen Verwertung von Holzaschen unter besonderer Berücksichtigung der Knickholzpotenziale Schleswig-Holsteins. Dissertation, University of Kiel

Van der Paauw F (1971) An effective water extraction method for the determination of plant-available soil phosphorus. Plant Soil 34:467–481

Van Ray B, Van Diest A (1979) Utilization of phosphate from sources by six plant species. Plant Soil 51:577–589

Vance ED, Mitchell CC (2000) Beneficial use of wood ash as an agricultural soil amendment: case studies from the United States forest products industry. In: Power JF, Dick WA (eds) Land application of agricultural, industrial and municipal by-products. SSSA, Madison, pp 567–582

Yusiharni BE (2001) An evaluation of chicken litter ash, wood ash and slag for use as lime and phosphate soil amendments. Master thesis, University of Mataram

Zhang F-S, Yamasaki S, Nanzyo M (2002) Waste ashes for use in agricultural production I. Liming effect, contents of plant nutrients and chemical characteristics of some metals. Sci Total Environ 284:215–225

Zhu Y-G, He Y-Q, Smith SE, Smith FA (2002) Buckwheat (*Fagopyrum esculentum* Moench) has high capacity to take up phosphorus (P) from a calcium (Ca)-bound source. Plant Soil 239:1–8

Chapter 3
Mixtures of Bottom Wood Ash and Meat and Bone Meal as NPK Fertilizer

Trond Knapp Haraldsen, Per Anker Pedersen, and Arne Grønlund

Abstract Bottom wood ash (BWA) contains high concentrations of Ca, Mg, K and P, whereas meat and bone meal (MBM) has been found to be a good N and P fertilizer. In a pot experiment on soils with low contents of readily available K and P, the effects of a mixture of MBM and BWA were compared with those of mineral NPK (21–4–10), MBM, MBM plus K and Mg mineral fertilizer, and MBM and different types of crushed K- or Mg-rich rock.

The mixture of MBM and BWA gave the highest yield of barley, at the same level as mineral for NPK, and significantly higher than for MBM alone. For the yield of spring wheat there was no significant difference between the treatments with BWA or other K-rich additives to MBM, MBM used alone and mineral NPK. Compared with the other treatments, the MBM and BWA mixture significantly increased the pH of the soils by 0.5 units. MBM plus BWA represents an interesting concept for development of recycled NPK fertilizer of organic origin.

3.1 Introduction

Many by-products from the food industry and waste products from industry and bioenergy plants contain high concentrations of plant nutrients. The degree of recycling of these plant nutrients differs, but there is an untapped potential for use of by-products and waste products as fertilizers. The bottlenecks for use are partly lack of knowledge of fertilization and liming effects, logistic problems, relatively cheap disposal of ash at landfills, heavy metal content and restrictions due to governmental regulations (see Chap. 1, Knapp and Insam 2011; Chap. 8,

T.K. Haraldsen (✉) and A. Grønlund
Soil and Environment Division, Norwegian Institute for Agricultural and Environmental Research, Frederik A. Dahls vei 20, 1432 Ås, Norway
e-mail: trond.haraldsen@bioforsk.no

P.A. Pedersen
Department of Plant and Environmental Science, Norwegian University of Life Sciences, P.O. Box 5003, 1432 Ås, Norway

H. Insam and B.A. Knapp (eds.), *Recycling of Biomass Ashes*,
DOI 10.1007/978-3-642-19354-5_3,

Saraber et al. 2011). In Norway the regulations for use of organic fertilizers and related material also apply for ash materials (Norwegian Ministry of Food and Agriculture 2003). These regulations are based on quality classes of heavy metals on a dry matter basis. Ash has increased concentrations of metals on a dry matter basis compared with the original organic material, and the regulations therefore strongly limit use of ash. In Norway most of the bioenergy plants collect bottom and fly ash in one container, giving a blend ash with higher concentrations of heavy metals than are allowed to be used on agricultural land.

Ash from bioenergy plants has mainly been considered as liming material because of its high content of Ca, although the Mg, K and P content may also be of importance. Owing to the strict regulations, ash has hardly been used in agriculture in Norway in recent years. Risse (2002) reviewed use of wood ash in agriculture in the USA, and referred to several investigations where liming with wood ash gave better growth than use of traditional limestone. The liming effects of ashes vary between 8 and 90% of the total neutralizing power of lime. Etiegni et al. (1991a, b) found that wood ash had positive effects as K fertilizer and as a liming agent, but the amounts should be restricted in accordance with the plants' need for K, and should not elevate the pH of the soil too much. The pH was the most important rate-limiting component. Saarelaa (1991) found that high application rates of wood ash supplied K in excess, and consequently decreased the Ca contents of plants. Mozaraffi et al. (2000) found that ash significantly increased K and decreased Mg concentrations in corn, and concluded that ash was a potentially good K source and a potential liming agent for acid soils. Different investigations on ash used as K or P fertilizer have shown different effects: almost the same of effect as wood ash as mineral K fertilizer (Erich 1991; Ohno 1992), whereas the P effects were better correlated with the citrate P than the total P in comparison with mineral P fertilizer (Erich 1991; Hansen 2004). Clapman and Zibelske (1992) found that plant uptake of Mg, K, Mn and P increased as the wood ash amendment increased. Meyers and Kopecky (1998) found land application of wood ash to be an environmentally safe alternative to landfilling, which may replace conventional limestone and K fertilizer for forage crop production.

Meat and bone meal (MBM) contains appreciable amounts of N, P and Ca, making it interesting as fertilizer for various crops. MBM has good effects as N fertilizer (Salomonsson et al. 1994, 1995; Jeng et al. 2004), and has a positive effect on baking performance of wheat (Fredriksson et al. 1997, 1998). Jeng et al. (2006) showed that the relative P efficiency of MBM was 40–50% compared with P in superphosphate (YARA P8) in experiments with cereals and rye grass, and Ylivainio et al. (2008) found MBM suitable as a long term P supply to perennial crops. Because use of MBM as feed to produce animals is banned in the European Union, the meat industry has focused on an alternative use of MBM. The interest in the use of MBM as fertilizer will increase if MBM is included in an organic mineral fertilizer which has a predictable NPK effect similar to a compound mineral NPK fertilizer. The annual production of MBM of category 3 material is about 30,000 t. MBM of category 3 material is allowed for use as fertilizer in Norway for all crops, except grassland which is used for grazing or mowing, but can also be used for grassland if it is mixed with other fertilizer materials (Norwegian Ministry of Food and Agriculture 2007). Combining animal bones, feathers and wood ash to make mineral NPK fertilizers was described by Chojnacka et al. (2006), but the fertilizers had high P content relative to N content.

The aim of this investigation was to test different mixtures of MBM and additives of mineral K, crushed K-rich or Mg-rich rock and bottom wood ash (BWA), and to compare the effects with those from the use of compound mineral NPK fertilizer (Yara Fullgjødsel® NPK 21–4–10) and an unfertilized control. The results should be used for development of organic mineral fertilizers based on MBM and BWA or other suitable K-rich material, which give a predictable NPK effect similar to that of mineral NPK. The investigation also aimed to find suitable types of BWA with low concentrations of heavy metals and high concentrations of K and Mg which satisfy the quality requirements for use on agricultural land in Norway (Norwegian Ministry of Food and Agriculture 2003).

3.2 Materials and Methods

3.2.1 Chemical Analysis

Total metals, P and S in crushed rock, bottom ash and plant material were extracted by 7 M $KHNO_3$ and determined by inductively coupled plasma optical emission spectrometry according to ISO 11885 (ISO 2007). The pH of the soil was determined in a solid–water suspension (1:2.5, v/v). Readily available P, K, Mg, and Ca in soil were determined by extraction with 0.1 M ammonium lactate and 0.4 M acetic acid (pH 3.75) in a solid-to-solution ratio of 1:20 (w/v) (Egnér et al. 1960). Nonexchangeable K was extracted by 1 M HNO_3 according to Pratt (1965). The particle size distribution of the soils was determined according to Elonen (1971). Total N was determined as Kjeldahl N as described by Bremner (1960).

3.2.2 Experimental Design

A pot experiment in a greenhouse was performed. Two soils were used; a sand (Elverum) and a sandy loam (Øksna) (Table 3.1). The soils were selected for this experiment because of the low content of readily available K, Mg and P which was found by Jeng et al. (2006). Although the soils were sampled at the same locations as the soils used in the experiments of Jeng et al. (2006), assuming low concentrations of readily available P and K, analyses of the soils selected for this experiment showed higher concentrations of readily available K (Table 3.2) than were found in the previous investigation [Elverum 7.6 mg readily available K $(100\ cm^3)^{-1}$, Øksna 8.6 mg readily available K $(100\ cm^3)^{-1}$]. According to Øgaard et al. (2002), no yield response to applied K can be expected if the concentration of readily available K exceeds 10 mg $(100\ cm^3)^{-1}$ (8 mg 100 g^{-1} and bulk density 1.25 mg m^{-3}).

Table 3.1 Texture of the soils used in pot experiments

Soil	Gravel (>2 mm) (percentage of sample)	Percentage of material <2 mm						
		Sand			Silt			Clay
		2–0.6 mm	0.6–0.2 mm	0.2–0.06 mm	0.06–0.02 mm	0.02–0.006 mm	0.006–0.002 mm	<0.002 mm
Øksna	0.2	0.4	8.6	47.0	28.4	9.7	2.3	3.6
Elverum	0.4	0.6	61.7	34.3	1.4	0.0	1.0	1.0

The experiment was designed to supply NPK similar to that supplied by use of the compound mineral NPK fertilizer Yara Fullgjødsel® 21–4–10. On the basis of the contents of N and P in MBM, and of K in different crushed rock powders and BWA (Table 3.3), the amounts of different components in the experimental design (Table 3.4) were calculated using the following assumptions for calculation of effective amounts of NPK:

1. The N effect of MBM was estimated as 80% of Kjeldahl N as equal to mineral N based on Jeng et al. (2004).
2. The P effect of MBM was estimated as 50% of total P equal to the effect of mineral P (Jeng et al. 2006).
3. The amount of K extracted by 1 M HNO_3 from crushed rock powder and by 7 M HNO_3 from BWA was estimated as plant available and equal to mineral K.

Although the total amount of NPK applied differed between the treatments, based on the assumptions for effective amounts of N, P and K, it was intended to obtain almost the same effect of NPK as mineral NPK (treatments 3–12).

The pot size was 7.5 l, and the height of the soil in the pots was 20 cm. The fertilizer level was based on normal fertilization recommendations for cereals in Norway (120 kg N ha^{-1}), and the concentration of plant nutrients in the pots should be at the same level as in a 20-cm-deep plough layer. All amounts of added fertilizer were calculated on a hectare to pot basis. There were three replicates. The crops used in the experiment were spring barley (*Hordeum vulgare* cv Kinnan) and spring wheat (*Triticum aestivum*), and the amounts of K applied (60 kg K ha^{-1}) were in line with normal fertilization recommendations for cereals in Norway. Seeding was performed in May with 30 seeds in each pot. After germination the weakest plants were removed, leaving 20 plants per pot. Unfortunately, barley seeds from two different batches were used, and uneven germination of barley was recorded owing to poor germination of the seeds from one batch. However, in the pots with fewer than 20 plants, the lack of plants was partly compensated for by an increased number of tillers. A mean of 19 ears per pot was found at harvest for both barley and wheat. Therefore, the effect of uneven germination was found not to have a significant influence on the yield.

The intended temperature in the greenhouse was 15°C at night and 20°C during the day, but on warm days with outdoor temperature above 20°C the temperature inside the greenhouse was somewhat higher than the outdoor temperature, reaching 30°C during some summer days. The pots were initially irrigated 3 days a week, but in

Table 3.2 Carbon and readily available plant nutrients in the soils used in the pot experiments

	pH	TOC (g 100 g^{-1} DM)	Na-AL (mg 100 cm^{-3})	K-AL (mg 100 cm^{-3})	Mg-AL (mg 100 cm^{-3})	Ca-AL (mg 100 cm^{-3})	P-AL (mg 100 cm^{-3})
Øksna	6.7	0.4	1.2	15.2	4.2	42.3	3.3
Elverum	6.9	<0.1	2.5	11.4	3.3	27.9	3.5

TOC *total organic carbon*; DM *dry matter*; Na-AL *readily available Na*; K-AL *readily available K*; Mg-AL *readily available Mg*; Ca-AL *readily available Ca*; P-AL *readily available P*

Table 3.3 Chemical properties of crushed rock, wood ash and meat and bone meal used in the pot experiments

Parameter	Crushed rock powder				Bottom wood ash	MBM Mosvik
	Altagro	Oxaal	Røyneberg	Olivin		
pH	9.1	8.3	8.7	9.4	13.0	6.2
DM, g $(100\ g)^{-1}$	–	–	–	–	–	95.2
Loss on ignition, g $(100\ g)^{-1}$ DM	–	–	–	–	–	75.2
TOC, g $(100\ g)^{-1}$ DM	–	–	–	–	–	29.4
Kjeldahl N, g $(100\ g)^{-1}$ DM	–	–	–	–	–	9.31
Total P, g $(100\ g)^{-1}$ DM	0.05	0.05	0.05	0.001	1.2	3.95
P-AL, g $(100\ g)^{-1}$ DM	0.0008	0.0006	0.0036	0.0008	–	–
Total K, g $(100\ g)^{-1}$ DM	1.41	0.52	1.44	0.02	4.3	0.51
K-HNO_3, g $(100\ g)^{-1}$ DM	0.90	0.45	0.74	0.02	–	–
K-AL, g $(100\ g)^{-1}$ DM	0.06	0.01	0.01	0.00	–	–
Total Ca, g $(100\ g)^{-1}$ DM	2.01	4.10	0.65	0.03	37.0	9.1
Total Mg, g $(100\ g)^{-1}$ DM	0.65	1.36	1.09	19.7	2.8	0.20
Total S, g kg^{-1} DM	0.0012	0.0099	0.0014	0.0001	0.029	5.31
Zn, mg kg^{-1} DM	54.8	55.2	267	23.0	354	133
Pb, mg kg^{-1} DM	<4.0	9.6	4.4	<4.0	6.3	<7.5
Ni, mg kg^{-1} DM	15.3	56.9	29.3	2,050	71.2	9.0
Cu, mg kg^{-1} DM	7.8	23.1	7.2	9.8	152	37.9
Cd, mg kg^{-1} DM	<0.4	<0.4	0.5	<0.4	0.4	<0.2
Cr, mg kg^{-1} DM	13.9	58.5	34.5	406	73.6	1.6
Mn, mg kg^{-1} DM	650	483	522	457	17,000	
Hg, mg kg^{-1} DM	–	–	–	–	–	<0.01

MBM meat and bone meal; *K-HNO_3* nonexchangeable K

warm periods irrigation was carried out almost daily to prevent drought. The irrigation caused no leaching of plant nutrients from the pots during the experiment.

The nutrient balances were calculated on the basis of effective amounts of P and K applied and uptake of P and K in wheat grain. In the experiment straw was not harvested and nutrient uptake in the straw is therefore not included in the calculations of nutrient balance. After the experiment had finished, soil samples from all treatments from both the wheat and the barley experiments with both soils were taken (0–20 cm), 48 samples in total. The samples consisted of nine subsamples, three from each pot. The results for the soil samples in Table 3.7 represent means for both soils and both crops.

Table 3.4 Application of different fertilizers in the pot experiment

Treatment	Fertilizer per hectare	Estimated fertilizer effect (kilograms per hectare first season)		
		N	P	K
1	No fertilization (control)	0	0	0
2	1,700 kg MBM	120	24	8
3	1,700 kg MBM + 220 kg kalimagnesia (K, Mg, S)	120	24	64
4	1,700 kg MBM (pellets) + 220 kg kalimagnesia (K, Mg, S)	120	24	64
5	1,700 kg MBM + 6,000 kg Altagro	120	24	62
6	1,700 kg MBM + 6,000 kg Altagro + 1,000 kg Olivin	120	24	62
7	1,700 kg MBM + 12,000 kg Oxaal	120	24	62
8	1,700 kg MBM + 12,000 kg Oxaal + 1,000 kg Olivin	120	24	62
9	1,700 kg MBM + 7,000 kg Røyneberg	120	24	61
10	1,700 kg MBM + 7,000 kg Røyneberg + 1,000 kg Olivin	120	24	61
11	580 kg Yara Fullgjødsel® NPK 21–4–10	120	21	56
12	1,700 kg MBM + 1,200 kg BWA	120	24	60

BWA bottom wood ash

3.2.3 *Statistical Analysis*

One-way analysis of variance was carried out. For multiple comparisons the Ryan–Einot–Gabriel–Welch (REGWQ) multiple range test was applied with a significance level of $P = 0.05$, and the means presented followed by the same letter are not statistically different.

3.3 Results

The mixture of MBM and BWA gave the highest barley yield (Table 3.5), which was significantly higher than for MBM alone. The barley yield of the MBM and different combinations of MBM and crushed rock was at the same level as for mineral NPK. All treatments with MBM or mineral NPK gave significantly higher yields of both barley and wheat compared with the unfertilized control (treatment 1). The yield of wheat was at the same level for mineral NPK and MBM with or without K addition (Table 3.5), but there was a significantly lower yield for MBM pellets (treatment 4) and MBM plus Altagro plus Olivin compared with MBM powder alone (treatment 2).

Different additions of K did not influence the concentrations of K in the wheat grain (Table 3.6). The unfertilized control had a significantly lower concentration of N and a higher concentration of P in wheat grain than obtained for treatment with MBM plus K, Mg and S, and mineral NPK gave a significantly lower P concentration than the unfertilized control. The concentrations of plant nutrients taken up in barley grain were not analyzed.

Table 3.5 Grain yield of spring barley and spring wheat (means followed by the same letter are not statistically significant, $P = 0.05$)

Treatment	Fertilizer type	Yield (g DM pot^{-1})	
		Barley	Wheat
1	Control, no fertilizer	3,00d	1.93c
2	MBM	7.36c	10.04a
3	MBM + K, Mg, S	7.88c	7.83ab
4	MBM (pellets) + K, Mg, S	8.78abc	5.2b
5	MBM + Altagro	7.70c	7.27ab
6	MBM + Altagro + Olivin	9.20abc	6.62b
7	MBM + Oxaal	7.33c	7.35ab
8	MBM + Oxaal + Olivin	8.07abc	7.89ab
9	MBM + Røyneberg	10.07abc	7.56ab
10	MBM + Røyneberg + Olivin	10.94ab	7.80ab
11	Mineral NPK	10.19ab	7.97ab
12	MBM + BWA	11.46a	7.67ab

Table 3.6 Estimated effect of NPK, uptake of NPK in wheat grain and nutrient balance (means followed by the same letter are not statistically significant, $P = 0.05$)

Fertilizer	Estimated NPK effect (mg per pot)			NPK uptake in grain (g 100 g^{-1} DM)			NPK uptake in grain (mg per pot)			NPK balance (mg per pot)		
	N	P	K	N	P	K	N	P	K	N	P	K
Control, no fertilizer	0	0	0	1.90b	0.41a	0.49a	37	8	9	−37	−8	−9
MBM	450	90	30	2.47ab	0.27b	0.51a	247	27	51	203	63	−21
MBM + K, Mg, S	450	90	240	2.73ab	0.25b	0.46a	214	20	36	236	70	204
MBM (pellets) + K, Mg, S	450	90	240	3.12a	0.30ab	0.48a	162	16	25	288	74	215
MBM + Altagro	450	90	230	2.95ab	0.31ab	0.50a	214	23	36	236	67	194
MBM + Altagro + Olivin	450	90	230	2.95ab	0.29ab	0.49a	195	19	32	255	71	198
MBM + Oxaal	450	90	230	2.91ab	0.29ab	0.47a	214	21	35	236	69	195
MBM + Oxaal + Olivin	450	90	230	2.87ab	0.31ab	0.50a	226	24	39	224	66	191
MBM + Røyneberg	450	90	230	2.87ab	0.29ab	0.49a	217	22	37	233	68	193
MBM + Røyneberg + Olivin	450	90	230	2.94ab	0.28ab	0.50a	229	22	39	221	68	191
Mineral NPK	450	80	210	2.87ab	0.23b	0.44a	229	18	35	221	62	175
MBM + BWA	450	90	230	2.84ab	0.32ab	0.54a	218	25	41	232	65	189

The efficiency of the N applied was high and at the same level for the treatments with MBM or mineral NPK, whereas relatively small amounts of the P and K applied were taken up in the wheat grain (Table 3.6). On the basis of the nutrient uptake in wheat, the unfertilized control had a negative balance for N, P and K, whereas addition of MBM alone caused a negative K balance (Table 3.6). All the other treatments had a positive balance of P and K. The MBM treatment lowered the amount of readily available K in the soil, but the difference from the other treatments was not statistically significant. There was no significant change in the level of nonexchangeable K ($KHNO_3$) for any of the treatments compared with the unfertilized control. MBM plus BWA gave a significantly increased amount of readily available P compared with use of MBM pellets plus K, Mg and S and the

Table 3.7 Mean pH and means of readily available plant nutrients in soil after harvest of cereals (means followed by the same letter are not statistically significant, $P = 0.05$)

Fertilizer	pH	P-AL (mg per pot)	K-AL (mg per pot)	K-HNO_3 (mg per pot)	Mg-AL (mg per pot)	Ca-AL (mgper pot)
Control, no fertilizer	7.20b	218b	803a	2,618a	315a	3,045ab
MBM	6.90b	348ab	593a	2,505a	270a	2,873b
MBM + K, Mg, S	6.93b	330ab	698a	2,880a	308a	3,165ab
MBM (pellets) + K, Mg, S	6.90b	236b	810a	2,693a	360a	2,873b
MBM + Altagro	6.95b	330ab	780a	2,835a	345a	3,435ab
MBM + Altagro + Olivin	7.13b	345ab	705a	2,723a	278a	3,045ab
MBM + Oxaal	7.15b	361ab	638a	2,670a	255a	3,780ab
MBM + Oxaal + Olivin	7.20b	335ab	728a	2,693a	263a	3,863ab
MBM + Røyneberg	7.05b	323ab	750a	2,820a	330a	3,285ab
MBM + Røyneberg + Olivin	6.98b	359ab	668a	2,895a	503a	3,345ab
Mineral NPK	7.03b	267ab	765a	2,880a	278a	2,535b
MBM + BWA	7.63a	386a	743a	2,595a	345a	5,220a

control. MBM plus BWA also increased the amount of readily available Ca in the soil significantly compared with the treatments with MBM, MBM pellets plus K, Mg and S and mineral NPK. The only treatment that significantly increased soil pH was the MBM plus BWA treatment. The pH increase was around 0.5 (Table 3.7).

The BWA used in this experiment had a high concentration of Ca relative to K (Ca-to-K ratio, 8.6; Table 3.3). To find ash with a stronger effect as a K fertilizer than liming material, analyses of the chemical properties of bottom ash from other plants were performed (Table 3.8). BWA Akershus (Table 3.8) had a high concentration of K and a low concentration of Ca (Ca-to-K ratio, 1.8) and low concentrations of heavy metals. The wood used originated from a timber terminal at Gardermoen, where bioenergy wood from a large district in eastern Norway is collected. The ash of cereal waste had high concentrations of P and K and low concentrations of heavy metals. The analyses indicated differences in the Ca-to-K ratio between ash of spruce and ash of pine (Table 3.8). The wood used at Reinsvoll was dominated by spruce and the ash had properties similar to those of the ash of pure spruce wood.

3.4 Discussion

Different K sources combined with MBM gave an increased yield of barley compared with the use of MBM alone, whereas supply of K in addition to MBM did not significantly influence the yield of wheat. The soils used in this experiment had higher concentrations of readily available K than planned. The soils used in this experiment were within the group of sandy soils with a low content of acid-soluble K (acid-soluble K minus readily available K, 8–24 mg 100 g^{-1}) as described by Øgaard

Table 3.8 Chemical properties of some bottom ash types from Norwegian bioenergy plants

Parameter	Bottom ash					
	Akershus	Spruce	Reinsvoll 1	Reinsvoll 2	Pine	Cereal waste
TOC, g (100 g)$^{-1}$ TS	0.1	0.7	1.1	0.7	6.6	1.3
Total P, g (100 g)$^{-1}$ DM	1.7	1.4	1.3	1.3	0.7	4.9
Total K, g (100 g)$^{-1}$ DM	7.7	4.3	3.2	7.2	4.3	9.0
Total Ca, g (100 g)$^{-1}$ DM	14.0	36.1	32.4	28.2	15.0	–
Total Mg, g (100 g)$^{-1}$ DM	1.7	2.5	3.9	4.3	1.7	3.0
Total S, g (100 g)$^{-1}$ DM	0.08	0.19	0.29	0.21	0.06	–
Zn, mg kg^{-1} DM	200	69.5	106	107	279	190
Pb, mg kg^{-1} DM	7.8	10.7	5.5	7.1	34	15
Ni, mg kg^{-1} DM	16	25.9	26.4	30.9	26	36
Cu, mg kg^{-1} DM	75	79.5	116	120	34	93
Cd, mg kg^{-1} DM	0.6	<0.4	0.8	2.1	0.26	<0.5
Cr, mg kg^{-1} DM	15	17.9	29.5	47.2	47	61
Mn, mg kg^{-1} DM	17,000	30,400	36,100	39,300	10,000	2,800

et al. (2002). Øgaard et al. (2002) found no yield response to applied K in perennial grass leys when the amount of readily available K exceeded 10 mg (100 cm^3)$^{-1}$ (8 mg 100 g^{-1} and bulk density 1.25 mg m^3). In the present experiment the readily available K level was not significantly lowered owing to only a small negative K balance in the MBM treatment (Table 3.6), and was clearly above the expected minimum readily available K level based on texture (Øgaard et al. 2002). Sufficient K in the soil is therefore the most probable reason for not finding a significant effect of K supply on K uptake in wheat grain. Using sand of the same origin as the Elverum sand in this study, Haraldsen et al. (2010) found that application of 80 and 160 kg N ha^{-1} in MBM alone and other organic N fertilizers with low K content gave significant K deficiency and reduced barley yield compared with the use of the same amounts of mineral NPK and a liquid anaerobic digestate based on source-separated household waste. That batch of sand had a considerably lower content of readily available K than the sand used in this study, and had a readily available K level close to the expected minimum level as described by Øgaard et al. (2002).

Although the effect of K supply was not large in this study, the mixture of MBM and BWA gave at least the same yield as mineral NPK or supply of K, Mg and S in addition to MBM. The concentrations of potential plant-available K in the crushed rock powder types was too low for making commercial fertilizer products, and the amounts used in this experiment did not influence K uptake and did not cause any change in the level of readily available K or nonexchangeable K in the soils. Combining N-rich waste (human urine) and wood ash gave more biomass of red beets than mineral fertilizer (Pradhan et al. 2010), and Kuba et al. (2008) as well as Bougnom et al. (2010, 2011; see Chap. 7) found positive effects on growth of mixing wood ash with compost. These examples indicate a potential of combining

N-rich and K-rich waste streams as fertilizer or soil amendments. Especially for use in organic cropping there is demand for recycled NPK fertilizers, which have predictable effects.

The Ca-to-K ratio of the BWA used in the present experiment was not optimal in the mixture with MBM, and the ash also had a higher concentration of Ni than allowed in materials that can be used as fertilizers for agricultural crops in Norway (maximum quality class II according to the Norwegian Ministry of Food and Agriculture 2003). BWA Akershus (Table 3.8) has a low Ca-to-K ratio and can be categorized as quality class I according to the Norwegian Ministry of Food and Agriculture (2003). Because a smaller amount of BWA Akershus than of the ash used in this study is needed to obtain the same NPK ratio, it is expected that a mixture of MBM and BWA Akershus will cause a smaller pH increase than the mixture of MBM and BWA in this study. The pH increase of 0.5 after a single application of 1,200 kg BWA ha^{-1} represented an estimated addition of about 700 kg CaO equiv ha^{-1}, and caused a significant increase in the amount of readily available Ca in the soil (Table 3.7). According to Franzefoss (2007) natural acidification (by leaching and acid precipitation) represents an annual demand of lime of 100–200 kg CaO ha^{-1} in eastern Norway and 200–400 kg CaO ha^{-1} in coastal areas in western and central Norway. A liming effect of 200–300 kg CaO equiv ha^{-1} will be suitable for a fertilizer that is to be used annually for cereals. The ash of cereal waste from a milling plant had interesting properties as a PK fertilizer, as the levels of heavy metals were lower than the limits for use on agricultural land (Norwegian Ministry of Food and Agriculture 2003). For further development of organic mixtures based on MBM plus BWA, ash of similar quality as BWA Akershus (Table 3.8) will be selected. The challenge is to find sufficient quantities of BWA with similar properties, in order to establish commercial production of organic NPK fertilizer by combination of waste streams.

3.5 Conclusions

A mixture of MBM and BWA with a low Ca-to-K ratio and lower concentrations of heavy metals than the limits in the governmental regulations for organic fertilizers may give a NPK fertilizer with reliable effects of all three major plant nutrients. Such a fertilizer, based on combining waste streams from society, will be more complete fertilizer product than the raw materials represent and is of special relevance and interest for organic cropping. The BWA used in the mixture with MBM in the present experiment did not have optimal chemical properties, and gave too high an increase in pH for annual use for cereals. Further investigations on optimization of mixtures of MBM and BWA are needed before such fertilizer is ready for commercial production.

Acknowledgements The pot experiment was a part of a project supported by the programme "ORIO-Organic Waste Products and Recycling of Resources" and Norsk Protein AS (grant no.

403). Ellen Zakariassen is thanked for skilful technical assistance and experimental work. Additional analyses of ash and further work on the fertilizer concept were carried out as a part of WP 1.4 "Residues upgrading and use" in CenBio, Bioenergy Innovation Centre (http://www.cenbio.no), which is supported by the Research Council of Norway, Norsk Protein AS and Akershus Energi AS.

References

Bougnom BP, Knapp BA, Etoa FX, Elhottová D, Němcová A, Insam H (2010) Designer composts for ameliorating acid tropical soils: effects on the soil microbiota. Appl Soil Ecol 45:319–324

Bougnom BP, Knapp BA, Etoa FX, Insam H (2011) Possible use of wood ash and compost for improving acid tropical soils. In: Insam H, Knapp BA (eds) Recycling of biomass ashes. Springer, Heidelberg, pp 87–106

Bremner JM (1960) Determination of nitrogen in the soil by the Kjeldahl method. J Soil Sci 55:1–33

Chojnacka K, Baranska M, Gorecka H, Gorecki H (2006) Utilization of animal bones, feathers and wood ash in the making of mineral fertilizers. Przem Chem 85:1256–1259

Clapman WM, Zibelske LM (1992) Wood ash as a liming amendment. Commun Soil Sci Plant Anal 23:1209–1227

Egnér H, Riehm H, Domingo WR (1960) Untersuchungen über die chemische Boden-analyse als Grundlage für den Beurteilung des Nährstoffzustandes der Böden. II. Chemische Extrations Methoden zur Phosphor und Kaliumbestimmung. K Lantbrukshögsk Annal 26:199–215

Elonen P (1971) Particle-size analysis of soils. Acta Agral Fenn 122:1–122

Erich MS (1991) Agronomic effectiveness of wood ash as a source of phosphorus and potassium. J Environ Qual 20:576–581

Etiegni L, Campbell AG, Mahler RL (1991a) Evaluation of wood ash disposal on agricultural land. 1. Potential as soil additive and liming agent. Commun Soil Sci Plant Anal 22:243–256

Etiegni L, Mahler RL, Campbell AG, Shafii B (1991b) Evaluation of wood ash disposal on agricultural land. 2. Potential toxic effects on plant growth. Commun Soil Sci Plant Anal 22:257–267

Franzefoss (2007) Kalkhåndboka,, version 21 Jul 2007. http://www.kalk.no/Files/bibliotek/Kalkhåndbok_rev.pdf. Accessed 26 Jul 2010

Fredriksson H, Salomonsson L, Salomonsson A-C (1997) Wheat cultivated with organic fertilizers and urea: baking performance and dough properties. Acta Agric Scand Sect B Soil Plant Sci 47:35–42

Fredriksson H, Salomonsson L, Andersson R, Salomonsson A-C (1998) Effects of protein and starch characteristics on the baking properties of wheat cultivated by different strategies with organic fertilizers and urea. Acta Agric Scand Sect B Soil Plant Sci 48:49–57

Hansen TM (2004) Separation og genanvendelse af aske fra biobrædselanlæg. Miljøprojekt no. 962 2004. Miljøministeriet, Miljøstyrelsen, p 161

Haraldsen TK, Andersen U, Krogstad T, Sørheim R (2010) Liquid digestate from anaerobic treatment of source separated household waste as fertilizer to barley. In: Lasardi K, Manios T, Bidlingmaier K, de Bertoldi M, Diaz L, Stentiford EI (eds) Proceedings of the 7th international conference ORBIT 2010, Grafima, Thessaloniki, Greece, 2010, pp 564–569

ISO (2007) International Standard ISO 11885 Water quality – Determination of selected elements by inductively coupled plasma optical emission spectrometry (ICP-OES). Second edition 2007-08-01, The International Organization for Standardization (ISO), p 40

Jeng A, Haraldsen TK, Vagstad N, Grønlund A, Tveitnes S (2004) Meat and bone meal as nitrogen fertilizer to cereals in Norway. Agric Food Sci 13:268–275

Jeng A, Haraldsen TK, Grønlund A, Pedersen PA (2006) Meat and bone meal as nitrogen and phosphorus fertilizer to cereals and rye grass. Nutr Cycl Agroecosyst 76:183–191

Knapp BA, Insam H (2011) Recycling of biomass ashes – current technologies and future research needs. In: Insam H, Knapp BA (eds) Recycling of biomass ashes. Springer, Heidelberg, pp 1–16

Kuba T, Tschöll A, Partl C, Meyer K, Insam H (2008) Wood ash admixture to organic wastes improves compost and its performance. Agric Ecosyst Environ 127:43–49

Meyers NL, Kopecky MJ (1998) Industrial wood ash as a soil amendment for crop production. TAPPI J 81:123–130

Mozaraffi M, Rosen CJ, Russelle MP, Nater EA (2000) Corn and soil response to application of ash generated from gasified alfalfa stems. Soil Sci 165:896–907

Norwegian Ministry of Food and Agriculture (2003) Forskrift om gjødselvarer mv. av organisk opphav. FOR 2003-07-04 nr 951. http://www.lovdata.no/for/sf/ld/xd-20030704-0951.html. Accessed 23 Apr 2010

Norwegian Ministry of Food and Agriculture (2007) Forskrift om animalske biprodukter som ikke er beregnet på konsum. FOR 2007-10-27 nr 1254. http://www.lovdata.no/cgi-wift/ldles?doc=/sf/sf/sf-20071027-1254.html. Accessed 23 Apr 2010

Øgaard AF, Krogstad T, Lunnan T (2002) Ability of some Norwegian soils to supply grass with potassium (K) – soil analyses as predictors of K supply from soil. Soil Use Manage 18:412–420

Ohno T (1992) Neutralization of soil acidity and release of phosphorus and potassium by wood ash. J Environ Qual 21:433–438

Pradhan SK, Holopainen JK, Weisell J, Heinonen-Tanski H (2010) Human urine and wood ash as plant nutrients for red beet (*Beta vulgaris*) cultivation. Impacts on yield quality. J Agric Food Chem 58:2034–2039

Pratt PF (1965) Potassium. In: Black CA (ed) Methods of soil analysis, part 2. Chemical and microbiological properties. American Society for Agronomy, Madison, pp 1023–1031

Risse M (2002) Best management practice for wood ash as agricultural soil amendment. The University of Georgia, College of Agricultural and Environmental Sciences, Athens, 1142

Saarelaa I (1991) Wood, bark, peat and coal ashes as liming agents and sources of calcium, magnesium, potassium and phosphorus. Ann Agric Fenn 30:375–388

Salomonsson L, Jonsson A, Salomonsson A-C, Nilsson G (1994) Effects of organic fertilizers and urea when applied to spring wheat. Acta Agric Scand Sect B Soil Plant Sci 44:170–178

Salomonsson L, Salomonsson A-C, Olofsson S, Jonsson A (1995) Effects of organic fertilizers and urea when applied to winter wheat. Acta Agric Scand Sect B Soil Plant Sci 45:171–180

Saraber A, Cuperus M, Pels J (2011) Ash from combustion of cacao residues for nutrient recycling: a case study. In: Insam H, Knapp BA (eds) Recycling of biomass ashes. Springer, Heidelberg, pp 107–120

Ylivainio K, Uusitalo R, Turtola E (2008) Meat bone meal and fox manure as P sources for ryegrass (*Lolium multiflorum*) grown on a limed soil. Nutr Cycl Agroecosyst 81:267–278

Chapter 4
Wood Ash Effects on Soil Fauna and Interactions with Carbohydrate Supply: A Minireview

Jouni K. Nieminen

Abstract Wood ash effects on soil animals in a boreal forest ecosystem are reviewed focusing on recent results on interactive effects of wood ash and organic amendments, and laboratory microcosms as a tool to understand soil food webs are discussed. Loose wood ash can reduce the populations of enchytraeids, collembolans and mites, but increase nematode populations particularly in experimental laboratory ecosystems with little or no primary production. Recent studies indicate that the repressive effect on enchytraeids depends on carbon availability. Carbohydrate supply seemed to alleviate the negative wood ash effect on enchytraeid body size and abundance. The fact that carbon alleviated wood ash effects on enchytraeids without any change in pH supports the view that wood ash effects on soil animals are partly indirect consequences of altered food resources. Experimental evidence suggests that the negative wood ash effect on enchytraeids is partly linked to increased bacteria-to-fungi ratio after wood ash application, and that this may be counteracted by carbohydrate addition.

4.1 Introduction

Only a fraction of the wood ash generated in power plants is recycled back to the forest ecosystem. For example, in Finland some 50% of the wood ash generated in energy production is utilized, and most of this is used for purposes other than forest fertilization (Finnish Forest Industries Federation 2008). Although the forest ecosystem would, in principle, benefit from recycling the nutrients back to the harvested sites, there are also problems such as short-term effects of elevated pH and levels of heavy metals on soil biota (Pitman 2006; see Chap. 1, Knapp and Insam 2011). In their review, Aronsson and Ekelund (2004) concluded that wood ash effects on soil fauna need to be investigated further.

J.K. Nieminen
Department of Biological and Environmental Science, University of Jyväskylä, 40014 Jyväskylä, Finland
e-mail: jouni.k.nieminen@jyu.fi

H. Insam and B.A. Knapp (eds.), *Recycling of Biomass Ashes*,
DOI 10.1007/978-3-642-19354-5_4,

Wood ash effects on soil animals have been investigated both in small-scale laboratory microcosms and in large-scale field experiments. Most laboratory experiments did not include any plants or included only one small tree seedling. Microcosms containing only heterotrophs are subsequently called heterotrophic microcosms, and microcosms including plants are called autotrophic microcosms. Further, loose wood ash was used in most studies, but today granulated ash is increasingly used as a forest fertilizer. Ash granules dissolve slowly in forest soil and increase the pH and the concentration of exchangeable cations slower than loose ash (Eriksson 1998). The properties of pure wood ash used in some studies are well known. The ash content of young and deciduous trees is higher than that of old and coniferous trees and the ash content of bark is manyfold that of stem wood (Hakkila and Kalaja 1983). Wood ash contains 1–6% phosphorus (P), 6–22% potassium (K), 19–33% calcium (Ca) and 2–5% magnesium (Mg), depending on the tree species and the component (Hakkila and Kalaja 1983). The Ca content of bark and branch ash is higher in relation to P, K and Mg than that of stem wood ash, and compared with other tree species, birch and alder ashes are P-rich (Hakkila and Kalaja 1983). The quality of ash obtained from power plants is more variable, depending on the fuel used (Korpijärvi et al. 2009). For example, in Finland the ash used in forest fertilization is typically a mixture of wood (logging residue, bark, sawdust) and peat ash, which contains less P, K, Ca and Mg but more aluminium and iron than wood ash (Hakkila and Kalaja 1983), and some nutrients are lost during the process. Consequently, the nutrient concentrations of ash from power plants are somewhat lower than those of pure wood ash. It is difficult to see whether differing wood ash effects were due to experimental conditions or ash properties. Specifically, tests of wood ash effects on soil animals at different levels of organic carbon availability were lacking until recently.

In this minireview I first briefly summarize the literature on wood ash effects on soil animals in general. Then I focus on recent laboratory microcosm experiments (plants excluded) and greenhouse experiments (grasses and conifer seedlings included) testing the effects of loose wood ash at different levels of carbon availability on enchytraeids and nematodes. Finally, the results are discussed focusing particularly on the use of laboratory microcosms as a research method.

4.2 Wood Ash Effects on Soil Animals

A summary of the effects of different ashes on forest soil animal groups detected in both field and laboratory experiments is given in Table 4.1. Loose wood ash was used in most experiments (Table 4.1). Because the dissolution rate of granulated wood ash is much lower than that of hardened or loose ash (Eriksson 1998; Nieminen et al. 2005), the ash effects on soil organisms also differ between ash types. The solubility of $CaCO_3$ formed during stabilization is 2 orders of magnitude lower than that of CaO and $Ca(OH)_2$, and, therefore, stabilized ash causes a much smaller pH shock in soil than loose ash (Steenari and Lindqvist 1997).

Table 4.1 Summary of wood ash effects on soil animals detected in field and laboratory experiments. When several treatment levels were applied, the upper limit is shown

References	Study type	Ash quality and quantity	Target organisms	Effect[a]
Haimi et al. (2000)	Field	Loose ash (<5 Mg ha^{-1})	Enchytraeids	–
			Microarthropods	(–)
Huhta (1984)	Field/ laboratory	Loose ash (<6.7 Mg ha^{-1}) + superphosphate (1 Mg ha^{-1})	Enchytraeids	–
Liiri et al. (2001)	Laboratory	Loose ash (5 Mg ha^{-1})	Enchytraeids	–
Liiri et al. (2002a)	Laboratory/ field	Loose ash (<5 Mg ha^{-1})	Collembolans	–
			Mesostigmatid mites	–
			Oribatid mites	–
Liiri et al. (2002b)	Field lysimeter	Loose ash (3 Mg ha^{-1})	Nematodes	+
			Enchytraeids	+
			Microarthropods	ns
Liiri et al. (2002c)	Laboratory	Loose ash (5 Mg ha^{-1})	Enchytraeids	–
			Nematodes	NS
Liiri et al. (2007)	Laboratory	Loose ash (5 Mg ha^{-1})	Enchytraeids	–
Liiri et al. (2007)	Laboratory	Loose ash (5 Mg ha^{-1})	Nematodes	NS (–)
Lundkvist (1998)	Field	Self-hardened/granulated ash (<8 Mg ha^{-1})	Enchytraeids	NS
Nieminen (2008a)	Laboratory	Loose ash (1 Mg ha^{-1})	Enchytraeids	–
Nieminen (2008a)	Laboratory	Loose ash (1 Mg ha^{-1})	Nematodes	NS
Nieminen (2009)	Greenhouse	Loose ash (0.5 Mg ha^{-1})	Enchytraeids	NS
Nieminen (2009)	Greenhouse	Loose ash (0.5 Mg ha^{-1})	Nematodes	+
Hyvönen and Huhta (1989)	Laboratory/ field	Loose ash (<7 Mg ha^{-1})	Nematodes	+
Vilkamaa and Huhta (1986)	Field	Loose ash (7 Mg ha^{-1})	Collembolans	±
Huhta et al. (1983)	Laboratory	Loose ash (1.75 Mg ha^{-1}) + superphosphate (0.25 Mg ha^{-1})	Nematodes, enchytraeids, collembolans, mites	–
Koskenniemi and Huhta (1986)	Field	Loose ash (7 Mg ha^{-1})	Oribatid mites	–
			Mesostigmatid mites	±

NS not significant
[a]+ and – indicate an increase and a reduction of abundance or biomass of broad taxonomic groups, weak effects in *parentheses*, ± indicates a species-dependent effect

Macroarthropods react slowly to wood ash application (Huhta et al. 1986). Although very high wood ash doses (7 Mg ha^{-1}) can cause negative effects on microarthropods (Koskenniemi and Huhta 1986), their populations and community structure tolerate moderate wood ash application (Liiri et al. 2002a, b). The numbers of some collembolans even increased after wood ash application in a field study (Vilkamaa and Huhta 1986). In contrast, 1–5 Mg loose wood ash per hectare, particularly if mixed with humus, has been shown to reduce the populations of the enchytraeid *Cognettia sphagnetorum* (Vejd.) both in many laboratory studies and in a field study (Huhta 1984; Haimi et al. 2000; Liiri et al. 2001, 2002c, 2007; Nieminen 2008a). A higher dose, however, was needed for a negative response in the field than in laboratory experiments (Haimi et al. 2000). In contrast to loose ash, hardened or granulated ash did not affect enchytraeid abundance, although a transient increase in the cadmium content of enchytraeid biomass was detected in one field study (Lundkvist 1998).

In contrast to its effects on enchytraeids and microarthropods, loose wood ash has been found to increase the total numbers of nematodes (Hyvönen and Huhta 1989; Liiri et al. 2002b; Nieminen 2009).

C. sphagnetorum is an omnivorous litter- and microbe-feeding oligochaetan worm dominating the mesofaunal community in boreal coniferous forests, and it is considered a keystone species in those ecosystems (Huhta et al. 1998). Because this species is adapted to acid soil (Bååth et al. 1980; Standen 1982; Abrahamsen 1983; Huhta 1984), the negative wood ash effects have been explained by the increase of soil pH after wood ash application (Aronsson and Ekelund 2004). Changing pH can affect soil animals indirectly by changing the community composition of microbes (Perkiömäki and Fritze 2002).

In addition to nutritional effects, wood ash can also modify enchytraeid populations. When Nieminen and Haimi (2010) transferred enchytraeids exposed to wood ash in the forest to laboratory microcosms containing unamended Norway spruce forest humus, they found that enchytraeid populations originating from wood-ash-treated soil propagated slower and mineralized less nitrogen (N) than populations from untreated control forest.

In summary, the most conspicuous wood ash effect on soil fauna is the reduction of enchytraeid populations after loose wood ash treatment, especially in small-scale laboratory experiments including small plants or no plants and, hence, very limited carbon flow to the soil. In contrast, up to 8 Mg granulated ash per hectare had no effect on enchytraeid abundance in the forest (Lundkvist 1998).

4.3 Experimental Microcosms

The functioning of ecological systems can be studied at small spatiotemporal scales in controlled laboratory microcosms. In particular, the microcosm method has been used as a tool to understand the functioning of decomposer food webs (Scheu 2002; Huhta 2007). Although the experimental approach has limitations such as limited community composition and restricted mobility of animals (Kampichler et al. 2001), idealized model systems help to reduce the natural variability and exclude variables which are considered beyond the scope of the research. The dynamics of decomposer food webs and nutrients in such systems are fairly well known. Decomposer population dynamics in laboratory microcosms containing no autotrophs, and hence no carbon input, consists of one growth phase and a subsequent decline phase (Nieminen 2008b). The length of the population growth phase depends on the initial population size (Nieminen 2002). For example, the biomass of fungal-feeding nematodes increased exponentially for 3.5 weeks in organic forest soil, then crashed and remained low for 16 weeks (Nieminen 2008b). In a laboratory experiment using larger pieces of forest floor, nematode populations crashed after 3 weeks, then collembolan and enchytraeid populations crashed (Huhta et al. 1983). In such a system lacking carbon input and nutrient uptake by plants, inorganic N accumulates in soil (Nieminen 2008b).

In such severely carbon limited systems, addition of organic carbon available to microbes increases microbial biomass, and reduces the N concentration in soil (Sparling and Williams 1986; Schmidt et al. 1997; Dunn et al. 2006). In an early experiment, glucose increased both the biomass of fluorescein diacetate active mycelium and yeasts as well as fungal-feeding nematodes in pine microcosms (Bååth et al. 1978). However, the soil properties were quite unnatural, for example the pH was high (above 6.8), and algae thrived in microcosms but enchytraeids were lacking (Bååth et al. 1978). In coniferous forest soil, the pH and moisture usually limit yeast and algal growth. Extra carbon input is also reflected in the biomasses of higher trophic levels of the soil decomposer food web. Addition of carboxymethyl cellulose to microcosms containing mineral and organic soil, needle litter and a Scots pine seedling increased the biomass of both saprotrophic fungi and hyphal-feeding nematodes (Nieminen and Setälä 2001). Although the biomass of filamentous fungi increases after cellulose addition, addition of labile carbon such as sucrose can enhance the growth of early successional (r-strategist) microbes such as bacteria (Moore-Kucera and Dick 2008; Nottingham et al. 2009) and *Zygomycota* fungi (Hanson et al. 2008). Nieminen (2010) found that a sucrose addition equalling 100 kg C ha^{-1} maintained a stable nematode population for one growing season, and an increasing enchytraeid population. Sucrose addition did not alter net N mineralization rates, indicating that N mineralization by increased animal populations exactly balanced the N immobilization in microbial biomass. When the carbon addition rate is increased above 100 kg C ha^{-1} in Norway spruce forest soil, soil animals, which have orders of magnitude lower growth rates than microbes, cannot consume all the extra microbial biomass in one growing season, and as a consequence, N is immobilized in microbial biomass.

4.4 Interactive Effects of Labile Carbon and Wood Ash

Two recent studies (Nieminen 2008b, 2009) focused on the interaction between wood ash and carbohydrate supply in microcosms. The enchytraeid *C. sphagnetorum* was chosen as the target organism because of its importance in boreal forest soil, and because studies have consistently shown that it is sensitive to wood ash. The abundance of microbial-feeding nematodes provided information on microbial production.

Nieminen (2008b) studied the interactive effects of sucrose and loose wood ash on enchytraeids and nematodes in organic Norway spruce forest soil from which enchytraeids had been extracted using the Baermann wet funnel technique (O'Connor 1957). After extraction, the humus was sieved, weighed into 20 cotton-plugged glass jars and 20 individuals of the enchytraeid *C. sphagnetorum* were returned and each microcosm was inoculated with microflora and microfauna. Nieminen (2008b) treated the microcosms with loose wood ash (480 mg per microcosm, 1,000 kg ha^{-1}) and sucrose (1.6 g per microcosm, 1.3 Mg C ha^{-1}) in a full factorial design.

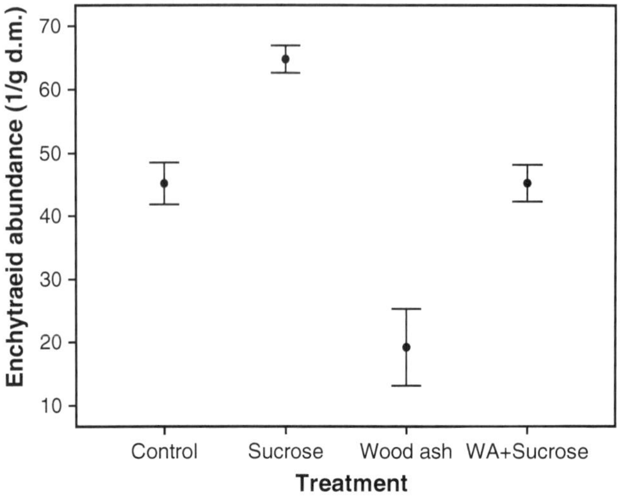

Fig. 4.1 Abundance of *Cognettia sphagnetorum* (Enchytraeidae) in organic Norway spruce forest soil amended with wood ash (*WA*) and sucrose after a 3-month incubation in laboratory microcosms (mean ± standard error, $n = 5$). (Data from Nieminen 2008a)

After a 3-month incubation at constant temperature, the dry matter content of organic soil was 22 ± 0.5% (mean ± standard error, $n = 20$) irrespective of treatment (Nieminen 2008b). Addition of wood ash increased the soil pH from 5.2 ± 0.1 to 6.9 ± 0.2 in both sucrose-amended and non-amended soil. Sucrose addition increased and wood ash addition reduced enchytraeid length. Sucrose addition increased nematode abundance by more than 100% on average. Wood ash addition alone decreased enchytraeid abundance compared with the control (Fig. 4.1).

Nieminen (2009) studied the interactive effects of three levels of solid sucrose (0, 0.88 or 8.8 g, equalling 0, 100 and 1,000 kg C ha^{-1}; the intermediate treatment was omitted in the original article) and two levels of birch ash (0 or 1.175 g per microcosm, equalling 500 kg ha^{-1}) in pots containing a layer of organic Norway spruce forest soil on mineral soil. In addition to one Norway spruce (*Picea abies* (L.) H. Karst. ssp. *abies*) seedling, these pots also contained the grasses *Deschampsia flexuosa* (L.) Trin. and *Calamagrostis epigejos* (L.) Roth and the experiment was continued in a greenhouse for one growing season (Nieminen 2009).

Wood ash increased the soil pH from 5.4 ± 0.06 to 5.7 ± 0.06 irrespective of carbon addition (Nieminen 2009). Sucrose increased the moisture of organic soil from 40 ± 0.02% to 57 ± 0.11% fresh mass in C1000 (analysis of variance, sucrose × ash $F_{2,24} = 4.9$, $p = 0.017$; control: simple effect of sucrose $F_{2,24} = 11.4$, $p < 0.001$), but this was partially counteracted by wood ash (moisture percentage 46 ± 0.05% in AC1000; $F_{1,24} = 7.5$, $p = 0.011$). Enchytraeids went close to extinction in control pots, but in C1000 treatments reached 1.8 times the initial density regardless of wood ash (sucrose $F_{2,24} = 10.5$, $p < 0.001$).

Enchytraeid length varied from 3.66 ± 0.28 mm in C1000 to 5.07 ± 1.46 mm in AC1000 ash-treated pots and the wood ash effect was significant. Wood ash increased nematode abundance (ash $F_{1,24} = 5.6$, $p = 0.027$). The nematode community in control pots consisted mainly of bacterial feeders (50% of individuals; in particular, *Rhabditis* sp.) and predators (*Mononchus* sp.), and all others amounted to less than 25% of individuals.

4.5 Discussion

4.5.1 Wood Ash Effects on Enchytraeids Depend on Carbon Availability

The most dramatic effects of wood ash and lime on enchytraeids have been reported in laboratory microcosms containing only a small tree seedling as a primary producer (Liiri et al. 2002c, 2007) or no plants (Pokarzhevskii and Persson 1995). It has long been realized that ash effects on decomposers are stronger in laboratory experiments than in the field, but the reason for this has been unknown (Huhta et al. 1986). In a recent experiment by Nieminen (2008b) the negative effects of wood ash on enchytraeid size and abundance were offset by sucrose addition without any change in pH, supporting the hypothesis that wood ash effects can be alleviated by relaxing carbon limitation to microbes.

Huhta et al. (1983) used quite large pieces (40 cm × 60 cm) of Scots pine forest soil. Nematode populations started to decline in control microcosms 3 weeks after the start of the experiments, and later collembolan and enchytraeid populations declined as well. A mixture of birch ash and superphosphate reduced the populations of enchytraeids, mites and later also collembolans as much as lime. Nematode populations were initially stimulated by the treatments, but the effect turned negative towards the end of the experiment. The experiment of Liiri et al. (2007) lasted for 1 year, and enchytraeids went close to extinction in untreated control pots as well as in ash-treated systems. In the experiment of Liiri et al. (2002c) enchytraeid biomass as well as the abundance of fungal-feeding nematodes declined even in unamended microcosms despite sufficient moisture, suggesting that the experimental ecosystems were severely resource limited. In the same way, the enchytraeid population declined in control microcosms in the experiment of Pokarzhevskii and Persson (1995). Nieminen and Setälä (2001) demonstrated carbon limitation in experimental microcosms: fungal-feeding nematodes propagated rapidly after cellulose addition to pine microcosms. Thus, carbon availability to microbes seems to be limiting decomposer activity in most laboratory microcosms, and consequently, resource limitation may have emphasized the effects of disturbances such as wood ash and lime observed in laboratory microcosms.

Carbon additions not only provide resources for microbes, they can also increase soil moisture (Szili-Kovács et al. 2007). Using a simple carbohydrate such as sucrose as the only carbon source has the advantage that it is easier to distinguish nutritional effects from other effects. The heterotrophic microcosms reported in Nieminen (2008b) were maintained relatively moist and because no carbon effect on moisture was detected at the end of the experiment, the carbon effects were probably nutritional in that study. In contrast, the topsoil in the greenhouse experiment (Nieminen 2009) was much drier at the end of the experiment. Since quite a large amount of solid sucrose was used (Nieminen 2009), a mulching effect was initially possible. In addition, sucrose is hygroscopic, but other mechanisms are also possible. In any case, the increased moisture in the C1000 sucrose treatment was critical for the persistence of enchytraeids (Nieminen 2009). Since enchytraeids are sensitive to drought, and consequently likely to suffer locally from the increasing frequency of extreme conditions such as drought (Lindberg et al. 2002), it can be suggested that more attention should be paid to the interaction of soil carbon and moisture.

The hierarchical nature of the soil food web is also worth noting. Even though low molecular weight carbon compounds are taken up and utilized on timescales ranging from seconds (Hill et al. 2008) to days (van Hees et al. 2002), the effects of labile carbon additions at higher trophic positions were still evident after one growing season.

4.5.2 Possible Mechanisms of Wood Ash Effects

The hypothesis that carbon amendment increased the ability of *C. sphagnetorum* to resist wood ash disturbance because of enhanced microbial production is supported by the large increase in the abundance of microbial-feeding nematodes (Nieminen 2008b), which is in accordance with the results of earlier field experiments. Bååth et al. (1995) found that wood ash application reduced the ratio of fungi to bacteria in pine forest soil. In a field study the ratio of fungal to bacterial phospholipid fatty acids was lower in forests treated with loose ash than in forests treated with hardened ash, although the difference from the control was not significant (Perkiömäki and Fritze 2002). Further, ash treatment was shown to alter the carbon utilization pattern of bacteria in a microcosm study, indicating changes in the bacterial community structure (Fritze et al. 2000). Although wood ash increases CO_2 evolution rates (Perkiömäki and Fritze 2002), Jokinen et al. (2006) concluded that not all microbial groups are equally stimulated by wood ash. In other words, by increasing the soil pH, wood ash application can favour fast-growing microbial species (Zimmermann and Frey 2002), which are not necessarily preferred food sources for dominant animals in forest soil (e.g. bacteria over fungi). An increase in microbial activity can also result from the bacterial decomposition of microbial residues. In particular, because the dominant enchytraeid species *C. sphagnetorum* is a litter feeder dependent on fungal activity, it is plausible that a shift towards the dominance of r-strategist bacteria has a negative influence on it. In accordance with this reasoning, Nieminen (2009) observed a tendency towards a lower proportion of bacterial-feeding nematodes in the sucrose treatment. We do not know whether sucrose increased the numbers of the same microbes that were negatively affected by wood ash, or if it increased other possible food sources for enchytraeids. On the basis of these and the present results, it seems that loose wood ash affects enchytraeids to some extent through a change in microbial community structure, and that sucrose counteracts this effect.

A moderate wood ash disturbance in the autotrophic experiment (Nieminen 2009) had little direct effect on decomposer animals. Since enchytraeids only persisted in sucrose-amended pots, the interaction of wood ash and carbon availability could not be tested. Although wood ash did not affect enchytraeid biomass, it did increase the size of individual worms, indicating that there were fewer but larger enchytraeids in the ash-treated microcosms. Because *C. sphagnetorum* reproduces asexually by fragmentation, a large individual size may be indicative of delayed reproduction. On the other hand, a higher wood ash application rate led to significantly smaller enchytraeid body size in the heterotrophic experiment (Nieminen 2008b), suggesting that when these organisms reproduce, they produce smaller offspring. This reasoning is supported by the data obtained by Nieminen and Haimi (2010), who found that enchytraeid populations originating from an ash-treated plot had a lower propagation rate (smaller fragmentation frequency) than populations from adjacent untreated control forest even when the animals were transferred to laboratory microcosms containing untreated organic soil. Also, the body size dynamics differed between populations with different disturbance histories. The overall length variation of the disturbed populations was smaller than that of the control populations, and, hence, the mean enchytraeid length could be either smaller or greater than that of control, depending on the sampling date (Nieminen and Haimi 2010).

4.5.3 Context-Dependent Wood Ash Effects on Nematodes

In accordance with earlier studies, wood ash increased nematode abundance (Hyvönen and Huhta 1989) in the autotrophic experiment (Nieminen 2009) but not in the heterotrophic experiment (Nieminen 2008b). Although nematodes were not identified, it is probable that their diversity was less in the simpler heterotrophic experiment, and this could partly explain their different response. Another major difference between the two experiments was that in the experiment without plants, soil moisture was kept optimal throughout the experiment, whereas in the pot experiment with plants, the organic soil dried considerably after grass harvest.

The increase of nematode numbers after loose wood ash application in field lysimeters with pine seedlings was mostly due to increased abundance of bacterial feeders, in other words, the ratio of bacterial feeders to fungal feeders increased (Liiri et al. 2002b). Liiri et al. (2007) found that loose wood ash reduced the abundance of fungal-feeding nematodes, but no effect on nematodes was evident in another microcosm study by Liiri et al. (2002c). The results are in broad accordance with an increased ratio of bacteria to fungi after ash application reported elsewhere (Bååth et al. 1995; Perkiömäki and Fritze 2002), although Liiri et al. (2002b) did not detect ash effects on the ratio of bacteria to fungi. This illustrates that given sufficient response time, microbe-feeding nematodes can reliably indicate ash effects on soil microbes. It remains unclear why nematodes are not always responsive to wood ash. One possible reason for this is the initial community composition. For example, in Liiri et al. (2002c), the abundance of fungal-feeding nematodes was low and highly variable at the beginning of the experiment, which obviously made it difficult to observe changes in the ratio of bacterial feeders to fungal feeders.

4.6 Conclusion

In summary, laboratory experiments have shown that wood ash effects on dominant enchytraeids depend on labile carbon availability to the decomposer food web. The fact that increased carbon availability alleviated wood ash effects on enchytraeids without changing the pH supports a view that wood ash effects on soil animals are partly indirect consequences of altered food resources. In other words, food limitation magnifies negative wood ash effects. Although the effects of loose wood ash are well known, knowledge of the effects of granulated ash and ashes amended with organic materials on soil organisms in boreal coniferous forests is still incomplete.

Acknowledgements The study was financially supported by the Runar Bäckström Foundation and the Maj and Tor Nessling Foundation. I appreciate the cooperation of the Jyväskylä Soil Ecology Group, help with Table 4.1 from J. Haimi, V. Huhta, M. Liiri and H. Setälä, and constructive comments on the manuscript by two anonymous reviewers.

References

Abrahamsen G (1983) Effects of lime and artificial acid rain on the enchytraeid (Oligochaeta) fauna in coniferous forest. Holarct Ecol 6:247–254

Aronsson KA, Ekelund NGA (2004) Biological effects of wood ash application to forest and aquatic ecosystems. J Environ Qual 33:1595–1605

Bååth E, Lohm U, Lundgren B, Rosswall T, Söderström B, Sohlenius B, Wirén A (1978) The effect of nitrogen and carbon supply on the development of soil organism populations and pine seedlings: a microcosm experiment. Oikos 31:153–163

Bååth E, Berg B, Lohm U, Lundgren B, Lundkvist H, Rosswall T, Söderström B, Wirén A (1980) Effects of experimental acidification and liming on soil organisms and decomposition in a Scots pine forest. Pedobiologia 20:85–100

Bååth E, Frostegård Å, Pennanen T, Fritze H (1995) Microbial community structure and pH response in relation to soil organic matter quality in wood-ash fertilized, clear-cut or burned coniferous forest soils. Soil Biol Biochem 27:229–240

Dunn RM, Mikola J, Bol R, Bardgett RD (2006) Influence of microbial activity on plant-microbial competition for organic and inorganic nitrogen. Plant Soil 289:321–334

Eriksson H (1998) Short-term effects of granulated wood ash on forest soil chemistry in SW and NE Sweden. Scand J For Res 2:43–55

Finnish Forest Industries Federation (2008) Forest industry's environmental statistics. Statistics for 2006. http://www.forestindustries.fi

Fritze H, Perkiömäki J, Saarela U, Katainen R, Tikka P, Yrjälä K, Karp M, Haimi J, Romantschuk M (2000) Effect of Cd-containing wood ash on the microflora of coniferous forest humus. FEMS Microbiol Ecol 32:43–51

Haimi J, Fritze H, Moilanen P (2000) Responses of soil decomposer animals to wood-ash fertilisation and burning in a coniferous forest stand. For Ecol Manage 129:53–61

Hakkila P, Kalaja H (1983) Puu – ja kuorituhkan palauttamisen tekniikka. Folia For 552:1–37

Hanson CA, Allison SD, Bradford MA, Wallenstein MD, Treseder KK (2008) Fungal taxa target different carbon sources in forest soil. Ecosystems 11:1157–1167

Hill PW, Farrar JF, Jones DL (2008) Decoupling of microbial glucose uptake and mineralization in soil. Soil Biol Biochem 40:616–624

Huhta V (1984) Response of *Cognettia sphagnetorum* (Enchytraeidae) to manipulation of pH and nutrient status in coniferous forest soil. Pedobiologia 27:245–260

Huhta V (2007) The role of soil fauna in ecosystems: a historical review. Pedobiologia 50:489–495

Huhta V, Hyvönen R, Koskenniemi A, Vilkamaa A (1983) Role of pH in the effect of fertilization on Nematoda, Oligochaeta and microarthropods. In: Lebrun PH, Andre HM, De Medts A, Grégoire-Wibo C, Wauthy G (eds) New trends in soil biology. Proceedings of the VIII international colloquium of soil zoology, Louvain-la-Neuve, Belgium. Dieu-Brichart, Ottignies-Louvain-la-Neuve, pp 61–73

Huhta V, Hyvönen R, Koskenniemi A, Vilkamaa P, Kaasalainen P, Sulander M (1986) Response of soil fauna to fertilization and manipulation of pH in coniferous forests. Acta For Fenn 195:1–30

Huhta V, Persson T, Setälä H (1998) Functional implications of soil fauna diversity in boreal forests. Appl Soil Ecol 10:277–288

Hyvönen R, Huhta V (1989) Effects of lime, ash and nitrogen fertilizers on nematode populations in Scots Pine forest soil. Pedobiologia 33:129–143

Jokinen HK, Kiikkilä O, Fritze H (2006) Exploring the mechanisms behind elevated microbial activity after wood ash application. Soil Biol Biochem 38:2285–2291

Kampichler C, Bruckner A, Kandeler E (2001) Use of enclosed model ecosystems in soil ecology: a bias towards laboratory research. Soil Biol Biochem 33:269–275

Knapp BA, Insam H (2011) Recycling of biomass ashes – current technologies and future research needs. In: Insam H, Knapp BA (eds) Recycling of biomass ashes. Springer, Heidelberg, pp 1–16

Korpijärvi K, Mroueh U-M, Merta E, Laine-Ylijoki J, Kivikoski H, Järvelä E, Wahlström M, Mäkelä E (2009) Energiantuotannon tuhkien jalostaminen maarakennuskäyttöön. VTT Tiedotteita – research notes 2499. http://www.vtt.fi/inf/pdf/tiedotteet/2009/T2499.pdf. Accessed 17 Aug 2010

Koskenniemi A, Huhta V (1986) Effects of fertilization and manipulation of pH on mite (Acari) populations of coniferous forest soil. Rev Écol Biol Sol 23:271–286

Liiri M, Setälä H, Haimi J, Pennanen T, Fritze H (2001) Influence of *Cognettia sphagnetorum* (Enchytraeidae) on birch growth and microbial activity, composition and biomass in soil with or without wood ash. Biol Fertil Soils 34:185–195

Liiri M, Haimi J, Setälä H (2002a) Community composition of soil microarthropods of acid forest soils as affected by wood ash application. Pedobiologia 46:108–124

Liiri M, Setälä H, Haimi J, Pennanen T, Fritze H (2002b) Soil processes are not influenced by the functional complexity of soil decomposer food webs under disturbance. Soil Biol Biochem 34:1009–1020

Liiri M, Ilmarinen K, Setälä H (2002c) The significance of *Cognettia sphagnetorum* (Enchytraeidae) on nitrogen availability and plant growth in wood ash-treated soil. Plant Soil 246:31–39

Liiri M, Ilmarinen K, Setälä H (2007) Variable impact of enchytraeid worms and ectomycorrhizal fungi on plant growth in raw humus soil treated with wood ash. Appl Soil Ecol 35:174–183

Lindberg N, Engtsson JB, Persson T (2002) Effects of experimental irrigation and drought on the composition and diversity of soil fauna in a coniferous stand. J Appl Ecol 39:924–936
Lundkvist H (1998) Wood ash effects on enchytraeid and earthworm abundance and enchytraeid cadmium content. Scand J For Res 2:86–95
Moore-Kucera J, Dick RP (2008) Application of ^{13}C-labelled litter and root materials for *in situ* decomposition studies using phospholipid fatty acids. Soil Biol Biochem 40:2485–2493
Nieminen JK (2002) No evidence of sensitivity to initial population size in a fungus-nematode soil food chain. Soil Biol Biochem 34:1821–1823
Nieminen JK (2008a) Labile carbon alleviates wood ash effects on soil fauna. Soil Biol Biochem 40:2908–2910
Nieminen JK (2008b) Linking food webs to ecosystem processes: piecewise linear models of soil microcosms. Ecol Model 217:87–94
Nieminen JK (2009) Combined effects of loose wood ash and carbon on inorganic N and P, key organisms and the growth of grasses and conifers in a pot experiment. Plant Soil 317:155–165
Nieminen JK (2010) Labile carbon effects on nematodes, enchytraeids and N mineralization in Norway spruce forest soil. Biol Fertil Soils 46:175–178. doi:10.1007/s00374-009-0407-6
Nieminen JK, Haimi J (2010) Body size and population dynamics of enchytraeids with different disturbance histories and nutrient dynamics. Basic Appl Ecol 11(7):638–644
Nieminen JK, Setälä H (2001) Influence of carbon and nutrient additions on a decomposer food chain and the growth of pine seedlings in microcosms. Appl Soil Ecol 17:189–197
Nieminen M, Piirainen S, Moilanen M (2005) Release of mineral nutrients and heavy metals from wood and peat ash fertilizers: Field studies in Finnish forest soils. Scand J For Res 20:146–153
Nottingham AT, Griffiths H, Chamberlain PM, Stott AW, Tanner EVJ (2009) Soil priming by sugar and leaf-litter substrates: a link to microbial groups. Appl Soil Ecol 42:183–190
O'Connor FB (1957) An ecological study of the enchytraeid worm population of a coniferous forest soil. Oikos 8:161–199
Perkiömäki J, Fritze H (2002) Short and long-term effects of wood ash on the boreal forest humus microbial community. Soil Biol Biochem 34:1343–1353
Pitman RM (2006) Wood ash use in forestry – a review of the environmental impacts. Forestry 79:563–588
Pokarzhevskii AD, Persson T (1995) Effects of oxalic acid and lime on the enchytraeid *Cognettia sphagnetorum* (VEJD) in mor humus. Water Air Soil Pollut 85:1045–1050
Scheu S (2002) The soil food web: structure and perspectives. Eur J Soil Biol 38:11–20
Schmidt IK, Michelsen A, Jonasson S (1997) Effects of labile carbon on nutrient partitioning between an arctic graminoid and microbes. Oecologia 112:557–565
Sparling GP, Williams BL (1986) Microbial biomass in organic soils: estimation of biomass C, and effect of glucose or cellulose amendments on the amounts of N and P released by fumigation. Soil Biol Biochem 18:507–513
Standen V (1982) Associations of Enchytraeidae (Oligochaeta) in experimentally fertilized grasslands. J Anim Ecol 51:501–522
Steenari B-M, Lindqvist O (1997) Stabilisation of biofuel ashes for recycling to forest soil. Biomass Bioenergy 13:39–50
Szili-Kovács T, Török K, Tilston EM, Hopkins DW (2007) Promoting microbial immobilization of soil nitrogen during restoration of abandoned agricultural fields by organic additions. Biol Fertil Soils 43:823–828
Van Hees PAW, Jones DL, Godbold DL (2002) Biodegradation of low molecular weight organic acids in coniferous forest podzolic soils. Soil Biol Biochem 34:1261–1272
Vilkamaa P, Huhta V (1986) Effects of fertilization and pH on communities of Collembola in pine forest soil. Ann Zool Fenn 23:167–174
Zimmermann S, Frey B (2002) Soil respiration and microbial properties in an acid forest soil: effects of wood ash. Soil Biol Biochem 34:1727–1737

Chapter 5
Characterization of Olive Waste Ashes as Fertilizers

Rogelio Nogales, Gabriel Delgado, Mar Quirantes, Manuel Romero, Esperanza Romero, and Eduarda Molina-Alcaide

Abstract Wet and dry olive cakes are the most important wastes generated when olive oil is produced. In recent years, both olive wastes have been incinerated to produce electricity, and thereby large amounts of fly and bottom ash are generated. In this study, physical, physicochemical, and chemical characteristics of olive waste ashes produced in Andalusian biomass power plants were analyzed to evaluate their suitability for agriculture. High variability among fly and bottom ashes may be ascribed to the origin of the olive waste and the combustion temperature. Waste olive ashes, which contained all particle sizes, showed high values of pH, salinity, water holding capacity, calcium carbonate equivalent, and P, K, Cu, and B contents. In contrast, moderate values were recorded for Ca, Mg, Zn, and Ni. Nitrogen is scarce in olive waste ashes; they thus can only be part of any fertilization strategy.

5.1 Introduction

The renewable energy consumption in the EU will increase from 11.6% in 2009 to 20% by 2020. Biomass is the main (61%) resource for the renewable energy consumed (10th EurObserv´ER 2010). In Spain, the contribution of renewable energy tp total gross domestic consumption in 2008 was 7.6%, of which biomass up 5.1 Mt (47% of total renewable energy, INE 2010).

Olive oil production is one of the most important industries in Mediterranean countries. In Spain, olive tree cultivation mainly occurs in Andalusia. In 2008, 600×10^3–900×10^3 t olive oil was produced in this region, using the two-phase centrifugation systems as common (90%) extraction technology (Fig. 5.1). This system generates huge amounts (between 2.5 and 3.5 Mt/year in Andalusia) of a

R. Nogales (✉), M. Quirantes, M. Romero, E. Romero, and E. Molina-Alcaide
Estación Experimental del Zaidín (CSIC), Profesor Albareda, 1, 18008 Granada, Spain
e-mail: rogelio.nogales@eez.csic.es

G. Delgado
Department of Edaphology, University of Granada, Campus Universitario de la Cartuja, 18071 Granada, Spain

H. Insam and B.A. Knapp (eds.), *Recycling of Biomass Ashes*,
DOI 10.1007/978-3-642-19354-5_5,

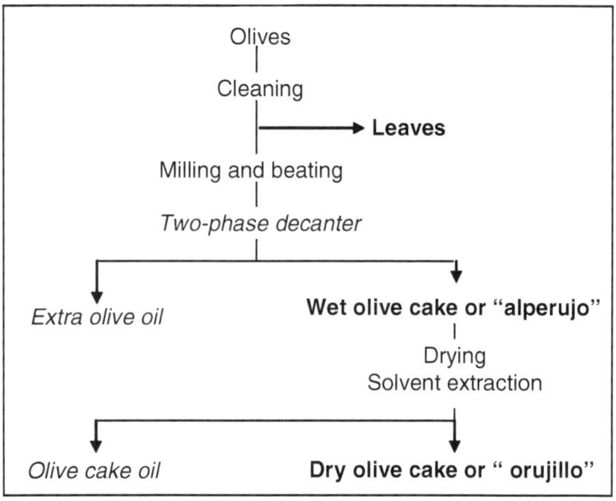

Fig. 5.1 The two-phase centrifugation system for olive oil extraction

waste called crude wet olive cake, or *alperujo*, which is composed of olive pulp, stones, and skins together with residual oil, and water added during the oil extraction process. *Alperujo* is an acidic, semihumid waste, rich in organic matter and potassium. After drying (less than 1% moisture), its low heating values ranges from 15 to18 MJ kg^{-1}. Part of this waste is dried and treated with solvents to obtain olive-cake oil and a waste called dry olive cake, or *orujillo*. The *orujillo* (0.6–0.9 Mt/year in Andalusia) has recalcitrant organic matter, high potassium content, and low heating values between 14.5 and 20 MJ kg^{-1} (Agencia Andaluza de Energía 1999; Alburquerque et al. 2004; Caputo et al. 2003; Nogales et al. 1998).

In recent years, both olive wastes are being used as fuel for electrical energy production. In Andalusia, ten biomass power plants have been established and produced 0.81 TWh (80% of the total renewable electricity generation in this region) in 2009. In general, the previously mentioned olive wastes, alone or mixed with other wastes, are burned in conventional boilers at 450°C. The heat released is used to heat water to turn a steam turbine, which generates green electric energy. In some biomass power plants, fluidized-bed combustors at 850°C are used as boilers. The combustion of olive wastes for energy production generates great amounts of fly ash and bottom ash (between 4 and 8% of the total burned olive wastes) as end waste. In general, this end waste is landfilled in sites adjacent to the biomass power plants. However, the use of landfills for ash disposal is expensive and is being discouraged by more stringent regulations and public opposition.

Numerous studies focused on chemical characteristics of ash produced by coal combustion or gasification (Ahmaruzzarman 2010; Hytonen 1998; Jala and Goyal 2006). In contrast, ash generated from biomass has received less attention and most of the research has focused on ash from woody biomass combustion (Demeyer et al. 2001; Kuba et al. 2008; Someshwar 1996; Vance 1996; see Chap. 6, Omil et al. 2011). Other ashes have been characterized, such as those from the pulp and paper industry (Naylor and Schmidt 1989; Muse and Mitchell 1995) and those produced by the incineration of municipal solid waste (Zhang et al. 2002) and biosolids (Benítez et al. 2001; Merino et al. 2005) or other agricultural wastes (Mozaffari et al. 2000). However, information is not available concerning the characteristics of ashes from olive waste combustion.

The aim of this study was to analyze the physical, physicochemical, and chemical characteristics of fly ash and bottom ash produced in Andalusian biomass power plants, which use olive wastes (wet and dry olive cakes) as fuel, to evaluate their suitability for use in agriculture.

5.2 Material and Methods

Representative samples of fly and bottom ashes from dry (*orujillo*) or wet (*alperujo*) olive cake, alone or mixed with other wastes were collected at different biomass power plants located in Andalusia (Spain, in 2008 and 2009. Table 5.1 shows the production characteristics and the types number of samples of olive waste ashes. The physical characteristics were determined for original samples, whereas the physicochemical and chemical analyses were carried out with ground (2-mm) samples using an agate mortar. Each analysis was replicated three times.

5.2.1 *Physical Analyses*

The particle size distribution was determined by sieving and sedimentation by the Robinson pipette method. The following particle fractions were separated: 0–50 μm, 50–250 μm, 250–2,000 μm, 2,000–5,000 μm, and more than 5,000 μm. Water retention at a matrix potential of 33 and 1,500 kPa was measured on a Richards pressure membrane.

Table 5.1 Production characteristics and types of olive waste ashes

Origin	Olive waste consumption (t year^{-1})	Type of incinerator	Combustion temperature (°C)	Type and number of ash samples analyzed
Wet olive cake	180,000	Moving grate	450	Fly ash (3)
Wet olive cake/ olive leaves	225,000	Moving grate	450	Fly ash (3) Bottom ash (3)
Wet olive cake/ olive leaves	53,000	Moving grate	450	Fly ash (3) Bottom ash (3)
Wet olive cake/ baggase	150,000	Fluidized bed	850	Fly ash (3) Bottom ash (3)
Dry olive cake	45,000	Fixed grate	450	Fly ash (3) Bottom ash (3)
Dry olive cake	104,000	Fixed grate	450	Fly ash (3) Bottom ash (3)

5.2.2 Physicochemical Analyses

The pH was measured using mixtures of ash and water (1:2.5) and the electrical conductivity was measured on a 1:5 ash–water extract. The cation exchange capacity (CEC) was determined by the ammonium acetate (pH 7) displacement method (Jackson 1970).

5.2.3 Chemical Analyses

The calcium carbonate equivalent (CCE) content was determined by analyzing the evolution of CO_2, after reaction with dilute hydrochloric acid, in a Bernard calcimeter (MAPA 1986). The total organic carbon content was determined by dichromate oxidation of the samples and subsequent titration with ferrous ammonium sulfate (Yeomans and Bremner 1989). The total Kjeldahl N was determined by semimicro-Kjeldahl digestion (Bremner and Mulvaney 1982). The total concentrations of P, K, Ca, Mg, Na, Fe, Mn, Cu, Zn, Cd, Ni, Pb, and B were analyzed by inductively coupled plasma mass spectrometry after digestion of samples in concentrated aqua regia (McGrath and Cunliffe 1985).

5.2.4 Statistical Analysis

Statistical analyses were carried out using the Statistical Package for Social Science (SPSS® Windows version 13.0) and Excel Statistics 2003 for Windows.

5.3 Results and Discussion

Figure 5.2 shows the particle size distribution for ashes from combustion of olive wastes. These particle sizes in ashes from olive wastes ranged from those for coarse sand to those for clay as stated by the USDA/FAO. As observed in other power plants that use coal or biomass as fuel, the particle size in fly ash is lower than in bottom ash. Sixty-five to 90% of particles in fly ash were less than 250 μm, whereas the percentage particles of that size was very low (1–25%) in the bottom ash. These particle sizes are similar to those observed in ashes from paper mill sludge (81.6% smaller than 250 μm) or wood (80% smaller than 1,000 μm) combustion as observed by others (Muse and Mitchell 1995; Etiegni and Campbell 1991). The combustion of olive wastes at 850°C significantly reduced the particle size of fly ash (43% smaller than 50 μm) compared with the combustion at 450°C (4–16% smaller than 50 μm). However, this effect did not occur in the bottom ash.

Water holding capacity is a physical parameter rarely analyzed in ashes from combustion of coal and biomass. However, it is well recognized that application of

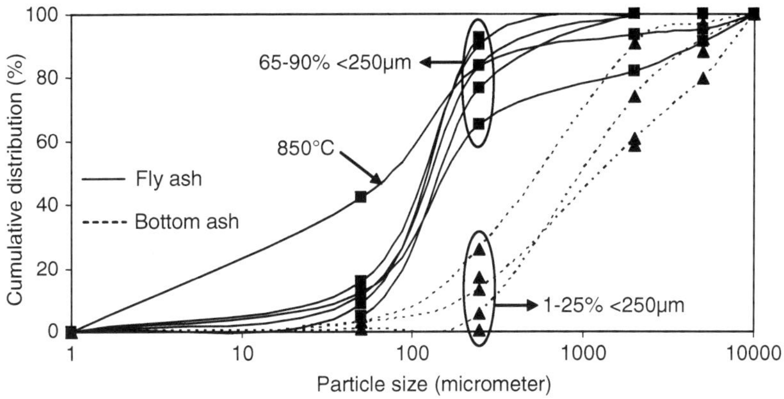

Fig. 5.2 Particle size distribution of ashes from olive waste combustion

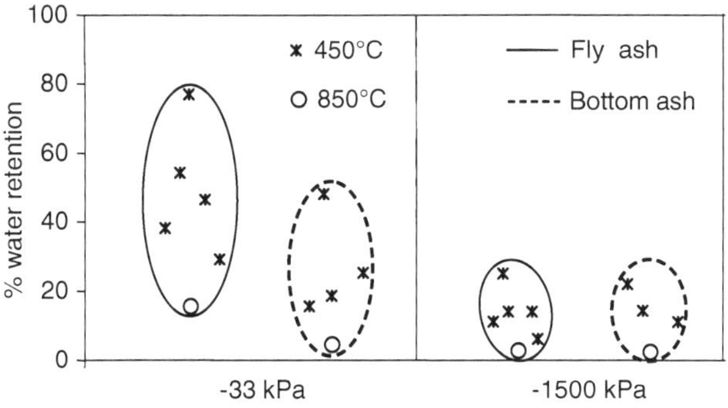

Fig. 5.3 Water retention at 33 and 1,500 kPa of ashes from olive waste combustion

fly ash to soil increases the water holding capacity at different matrix potentials and, in addition, the plant-available water of soils (Campbell et al. 1983; Pathan et al. 2003; Sharma et al. 1990). Fly ash shows a higher water holding capacity at −33 kPa (field capacity) than bottom ash from olive waste combustion, although both types of ashes showed a high variability (Fig. 5.3) induced by the temperature of combustion (higher water holding capacity in ashes from olive wastes burned at 450°C) and also by the type of olive waste burned (significant higher values for wet olive cake mixed with leaves). In contrast, at −1,500 kPa (permanent wilting point), the variability decreased, and there were no significant differences between fly ash and bottom ash. Finally, the water capacity (or the difference between the water content at the field capacity and that at the permanent wilting point of the ashes) was significantly higher in fly ashes from combustion of olive wastes than in the bottom ashes.

As observed in ashes from wood or other biomass sources, ashes from combustion of olive wastes showed strongly alkaline pH (between 9.9 and 13.7; Table 5.2).

Table 5.2 pH, calcium carbonate equivalent (*CCE*), effective neutralizing value (*ENV*), electrical conductivity (*EC*), cation exchange capacity (*CEC*) and organic carbon in ashes from olive wastes combustion

	Fly ashes ($N = 18$)				Bottom ashes ($N = 15$)			
	450°C		850°C		450°C		850°C	
	Mean	Range	Mean	Range	Mean	Range	Mean	Range
pH	11.6	9.9–13.3	12.2	12–12.4	13.3	13–13.7	11.8	11.6–12
CCE (g kg^{-1})	269	111–453	46	41–50	174	116–298	24	22–26
ENV (%)	24	9.6–41	4.3	3.8–4.6	6.25	4.6–9.4	2.4	2–2.6
EC (mS cm^{-1})	110	59–139	78	72–81	45	24–66	27	24–30
CEC (cmol kg^{-1})	14.4	7.2–20	11	9.9–11.6	16.9	11–26	2.5	1.5–3.5
Organic C (g kg^{-1})	6.9	4.3–14.4	7.3	6.8–7.8	19.4	12–27	1.5	1–2

In general, the pH values of bottom ashes are slightly higher than those of fly ashes, although this difference was not significant owing to the high variability of pH recorded between the different samples assayed. The alkalinity depended, mainly, on the carbonate, bicarbonate, and hydroxide contents in the different ashes analyzed.

The (CCE) of the ashes, a parameter closely related to the pH, ranged from 22 to 453 g kg^{-1} (Table 5.2). In general, the values were lower than those found for wood, paper, and pulp ashes. Vance (1996) reported an average CCE of 481 g kg^{-1}, ranging between 132 and 924 g kg^{-1}, for 18 different wood-fired boiler-ash samples. The high variability recorded in our study is due to the type of olive waste burned, the type of ash (fly or bottom) produced, and the combustion temperature. In general, ashes from dry olive cake combustion had significantly higher CCE than those from wet olive cake. In addition, the CCE values were higher in fly ashes than in bottom ashes. The CCE decreased sharply with increased temperature of combustion. According to Etiegni and Campbell (1991), carbonates and bicarbonates predominate in ashes from combustion at 500°C, whereas oxides are prevalent in those from combustion above 1,000°C. In addition, the CCE can also change during ash storage and under varying environmental conditions such as divergent CO_2 amounts and moisture. Oxides can be hydrated to form hydroxides, which can subsequently react with CO_2 to form carbonates (Meiwes 1995).

The effective neutralizing value ranged from 3.8 to 41%, being appreciably higher in fly ash from combustion of olive wastes at 450°C (Table 5.2). This quality index, based on both purity and fineness, is widely used to express the effectiveness of a liming material for neutralizing soil acidity. Purity is measured as the CCE, whereas the fineness determines the reactivity or efficiency of the material. Therefore, only fly ashes from olive wastes could be considered as an acceptable liming agent to correct the pH of acidic soils.

Olive waste ashes had variable and high electrical conductivity (24–139 mS cm^{-1}) (Table 5.1), generally higher than those for ashes from coal and wood combustion (Kuba et al. 2008; Pathan et al. 2003). Comparatively, the electrical conductivities were significantly higher in fly ash obtained by combustion at 450°C. The high electrical conductivity, which reflects the concentration of total dissolved electrolytes in ashes, implies that a significant fraction of nutrients and metals present in ashes are dissolved basic salts (Nurmesniemi et al. 2005), which would have arisen during the combustion of olive wastes. The high electrical conductivity recorded, particularly in fly ash, could have an adverse effect on crops, particularly vegetables, if high doses of this type of ash are applied to the soil (Maas 1990).

The total CEC is another index of fertility rarely analyzed in ashes. Values of the CEC between 2.3 and 15.4 cmol kg^{-1} have been recorded in fly ashes from coal combustion (Pathan et al. 2003). Similar values (1.5–26 cmol kg^{-1}) were observed in ashes from combustion of olive wastes, with no differences between fly ash and bottom ash (Table 5.2), although an increased combustion temperature diminished significantly the CEC of both types of ashes. Although the CEC is directly related to the texture and organic matter content in soils, in the ashes assayed no correlation between the CEC and particle size and/or organic carbon content was found.

Although organic carbon is generally oxidized and transformed into gaseous constituents during combustion, its content in ashes is quite variable (1–14.4 g kg^{-1}, Table 5.2), indicating the presence of unburned organic matter, as a consequence of incomplete combustion of the olive wastes in the conventional boilers or fluidized-bed combustors. The presence of unburned carbon in the olive waste ashes could enhance their adsorbent properties. This has also been observed in other types of ashes, particularly those produced by the combustion of paper mill bark (Someshwar 1996; Muse and Mitchell 1995). Bottom ashes had higher levels of organic carbon (about 3 times) than fly ashes. Increase of the combustion temperature scarcely affected the organic carbon content in fly ash, whereas it decreased sharply in the bottom ashes.

Information concerning macronutrients and Na concentrations in the fly and bottom ash from combustion of olive wastes is shown in Fig. 5.4. High variability in

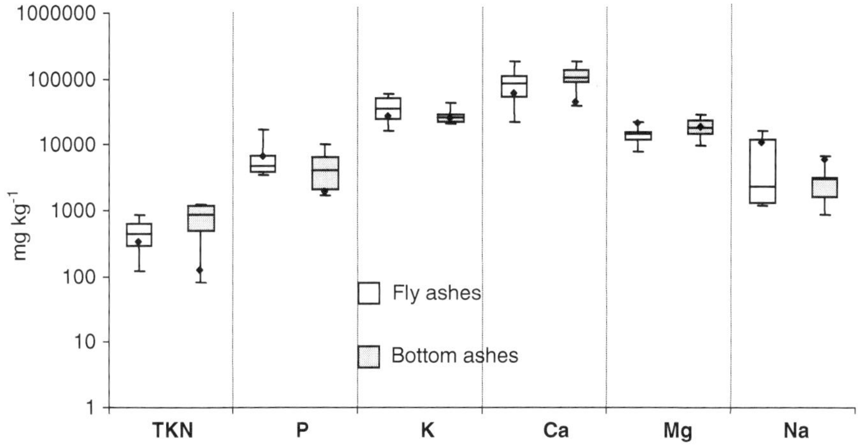

Fig. 5.4 Boxplots showing macronutrient and sodium contents in ashes from olive waste combustion. *Boxes* represent the interquartile range, with a *horizontal line* for the median, an d the *whiskers* over and under the boxes represent the 90 and 10% level data. The *black points* indicate the average value of each element in ashes from olive waste combustion at 850°C *TKN* total Kjeldahl nitrogen

all the elements was observed. A low N content was recorded, which may be due to the loss of this nutrient as gaseous N during the combustion of olive wastes. Significantly lower N content was found in ashes from combustion at 850°C than in those from combustion at 450°C. Calcium, K, Mg, and P were the major macronutrients in those ashes, and were found, mainly, in the form of carbonates owing to the high combustion temperature. In general, the P and K concentrations were higher in fly ash than in bottom ash, whereas the Ca and Mg concentrations were higher in bottom ashes. An increased combustion temperature promoted decreasing K and Ca concentrations and increasing Na concentration in both types of ashes from olive wastes. Etiegni and Campbell (1991) showed that, for the most part, the metal content in ashes increased with temperature, but the K, Na, and Zn content decreased, probably owing to the low boiling and decomposition points of their carbonates and oxides. In addition, volatilization of K has been observed with combustion temperatures varying from 800 to 900°C, resulting in losses of 63 and 90% of this nutrient (Naylor and Schmidt 1989).

As has been observed in other studies, macronutrients and Na concentrations in ashes depend on the type of biomass used as fuel, the type of incineration, the combustion temperature, and the type of ash (bottom or fly). The ashes studied in the present work had K concentration between 14 and 58 g kg^{-1}, higher than values observed in wood ashes (Etiegni et al. 1991; Kuba et al. 2008). This is due to the high K content characteristic of olive wastes (Alburquerque et al. 2004; Nogales et al. 1998; Melgar et al. 2009). Although an important fraction of the K contained in ashes should form insoluble complexes as a consequence of the high temperatures reached during the combustion of olive wastes (Lewis and Schmidt 1986), another fraction should be soluble and, as a consequence, should be rapidly available in the soil and extracted by crops (Ulery et al. 1993; Nogales et al. 2006). Phosphorus (3.5–17 g kg^{-1}), Ca (22–181 g kg^{-1}), Mg (8–22 g kg^{-1}), and Na (1–16 g kg^{-1}) concentrations in the ashes from olive wastes were similar to those from wood, paper, pulp, and coal (Huang et al. 1992; Jala and Goyal 2006; Ohno and Erich 1993). Therefore, olive waste ashes may be a direct source of P, K, Ca and Mg for soils, although the Ca and Mg contents in those ashes were less than those in liming agents currently used in agriculture.

A high variability was observed for trace element contents in ashes from olive wastes (Fig. 5.5). Fe was the most abundant element (1,519–26,240 mg kg^{-1}), whereas concentrations of Mn (71–400 mg kg^{-1}), Cu (92–797 mg kg^{-1}), Zn (91–622 mg kg^{-1}), and B (128–544 mg kg^{-1}) were more similar among most samples. Nickel concentrations (12–54 mg kg^{-1}) were lower, and Pb (3–70 mg kg^{-1}) was only detected in fly ashes and not in bottom ashes. Compared with ashes from coal and wood, the Cu and B contents in those from olive wastes were higher, whereas the contents of other trace elements were lower (Huang et al. 1992; Jala and Goyal 2006; Pitman 2006). However, owing to the broad ranges, comparisons are difficult for these trace elements. Compared with other liming agents, B and Cu concentrations were greater in ashes from olive wastes. The contents of Cd, Co, and Cr were below the detection limits (less than 0.02 mg kg^{-1})

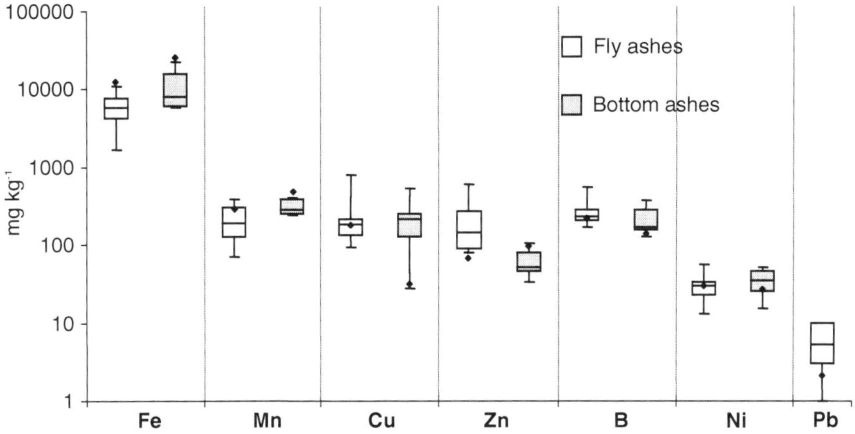

Fig. 5.5 Boxplots showing trace elements in ashes from olive waste combustion. *Boxes* represents the interquartile range, with a *horizontal line* for the median, and the *whiskers* over and under the boxes represent the 90 and 10% level data. The *black points* indicate the average value of each element in ashes from olive waste combustion at 850°C

in the olive waste ashes with the analytical procedure used. That could be due to both the low content of these trace elements in the olive wastes and possible losses during combustion (Belevi and Moench 2000).

It is generally accepted that the concentration of trace elements is significantly higher in fly ash than in bottom ash (Obernberger et al. 1997; Obernberger and Supancic 2009). Vaporization of metals taking place during the combustion processes is followed by condensation on cooler surfaces away from the heating zone and results in high heavy metal concentrations in fly ash (Hakkila 1989). However, in our study this was only observed for Zn and Pb. The concentrations of other trace elements did not decrease as was observed for B, Cu, Fe, and Mn.

Similar to macronutrients, with the exception of N, olive waste ashes contain high amounts of essential trace elements; therefore, these ashes could be used as soil amendments to increase the availability of those elements in the soil. The contents of heavy metals as Pb, Ni, Cd, Cr, and Co were low or negligible, so the use of the olive waste ashes as fertilizers would not constitute a risk for accumulation of those elements in soils and plants. Although in Spain there are no guidelines for the appropriate application of biomass ashes on agricultural and forest lands, most of the trace element contents in the olive waste ashes comply with the limits set in the guidelines of other countries, such as Austria, Germany, Denmark, Sweden, and Finland (Haglund and Expert group 2008; Obernberger and Supancic 2009). The only exception was Cu, which reached values varying from 506 to797 mg kg^{-1} in ashes from wet olive combustion mixed with leaves. Those values were higher than the maximum values allowed in the above-mentioned guidelines.

5.4 Conclusions

Energy production from olive wastes generates great amounts of fly ash and bottom ash. The characteristics of these ashes were examined in this study. High physical, physicochemical, and chemical variability in both fly and bottom ashes from olive wastes were observed. This variability may be due to factors such as the origin of the olive waste, the type of incinerator, the combustion temperature, and the type of ash (fly or bottom). Both fly and bottom ashes showed high values of pH, salinity, CCEs, and P, K, Cu, and B contents. In contrast, moderate or low values were recorded for Ca, Mg, Zn, and Ni. The contents of Cd, Co, and Cr were negligible. Fly ashes had a smaller particle size than bottom ashes, which resulted in higher neutralizing values in fly ash than in bottom ash. Therefore, fly ash from olive wastes may be used as soil amendment for liming purposes and potentially as forest fertilizer. Since N is scarce in fly ash from olive wastes, care has to be taken to provide supplemental N. On th basis of their lower fertilizer value, bottom ashes from olive wastes may be used for alternative purposes such as soil remediation or in industrial applications such as road and civil construction.

Acknowledgements This study was financed by the Junta de Andalucia, Spain, through project P07-RNM-2746. M.Q. and M.R. thank the Junta de Andalucia and CSIC, respectively, for their doctoral grants. We thank the biomass power plants of Agroenergetica de Baena, Vetejar, Agroeneregetica de Algodonales, El Tejar Autogeneración, and Planta de Biomasa La Loma for providing the fly and bottom ash from olive wastes used in this study.

References

10th EurObserv´ER report (2010) The state of renewable energies in Europe. Paris, France. Available via EurObserv´ER. http://www.eurobserv-er.org/pdf/barobilan10.pdf. Accessed 4 April 2011

Agencia Andaluza de Energía (1999) Potencial y aprovechamiento energetico de la biomasa de olivaren Andalucia. http://www.agenciaandaluzadelaenergia.es/agenciadelaenergia/portal/com/bin/contenidos/publicaciones/aprovechamiento_energetico/1130059713839_potencial_y_aprovechamiento.pdf. Accessed 11 May 2010

Ahmaruzzarman M (2010) A review on the utilization of fly ash. Prog Energy Combust Sci 36:327–363

Alburquerque JA, Gonzalvez J, Garcia D, Cegarra J (2004) Agrochemical characterisation of "alperujo", a solid by-product of the two-phase centrifugation method for olive oil extraction. Bioresour Technol 91:195–200

Belevi H, Moench H (2000) Factors determining the element behaviour in municipal solid waste incinerators. 1. Field study. Environ Sci Technol 34:2501–2506

Benítez E, Romero E, Gómez M, Gallardo Lara F, Nogales R (2001) Biosolids and biosolids ash as sources of trace elements in the plant soil system. Water Air Soil Pollut 132:75–87

Bremner JM, Mulvaney CS (1982) Nitrogen-total. In: Page AL, Miller RH, Keeney DR (eds) Methods of soil analysis, part 2. American Society for Agronomy, Madison, pp 594–624

Campbell DJ, Fox WE, Aitken RL, Bell LC (1983) Physical characteristics of sands amended with fly ash. Aust J Soil Res 21:147–154

Caputo AC, Scacchia F, Pelagagge PM (2003) Disposal of by-products in olive oil industry: waste-to-energy solutions. Appl Therm Eng 23:197–214
Demeyer A, Voundi JC, Verloo MG (2001) Characteristics of wood ash and influence on soil properties and nutrient uptake: an overview. Bioresour Technol 77:287–295
Etiegni L, Campbell AG (1991) Physical and chemical characteristics of wood ash. Bioresour Technol 37:173–178
Etiegni L, Campbell AG, Mahler RL (1991) Evaluation of wood ash disposal on agricultural land. I. Potential as soil additive liming agent. Commun Soil Sci Plant Anal 22:243–256
Haglund N, Expert group (2008) Guideline for classification of ash from solid biofuels and peat utilised for recycling and fertilizing in forestry and agriculture. NT technical 613. Nordic Innovation Centre, Oslo, Norway. Available via Nordic Innovation Centre. http://www.nordic innovation.net/nordtestfiler/rep613.pdf. Accessed 14 Jun 2010
Hakkila P (1989) Utilisation of residual forest biomass, Springer series in wood science. Springer, Berlin
Huang H, Campbell AG, Folk R, Mahler RL (1992) Wood ash as a soil additive and liming agent for wheat. Field studies. Commun Soil Sci Plant Anal 23:25–33
Hytonen J (1998) Effect of peat ash fertilization on the nutrient status and biomass production of short-rotation willow on cut-away peatland area. Biomass Bioenergy 15:83–92
INE (2010). Anuario Estadístico de España. Instituto Nacional de Estadística. Madrid. Available via INE. http://www.ine.es/prodyser/pubweb/anuario10/anu10_13indus.pdf. Accessed 5 April 2011
Jackson ML (1970) Análisis químico de suelos. Omega, Barcelona
Jala S, Goyal D (2006) Fly ash as a soil ameliorant for improving crop production – a review. Bioresour Technol 97:1136–1147
Kuba T, Tschöll A, Partl C, Meyer K, Insam H (2008) Wood ash admixture to organic wastes improves compost and its performance. Agric Ecosyst Environ 127:43–49
Lewis MN, Schmidt E (1986) Agricultural use of wood ash as a a fertilizer and liming material: field trials. TAPPI J 69:114–119
Maas EV (1990) Crop salt tolerance. In: Tanji KK (ed) Agricultural salinity assessment and management. ASCE manuals and reports on engineering 71. American Society of Civil Engineering, New York, pp 262–304
MAPA (1986) Métodos oficiales de análisis. Tomo III. Plantas, productos organicos, fertilizantes, suelos, agua, productos fitosanitarios y fertilizantes organicos. Publicaciones del Ministerio de Agricultura, Pesca y Alimentacion, Madrid
McGrath SP, Cunliffe CH (1985) A simplified meted for the extraction of the metals Fe, Zn, Cu, Ni, Cd, Pb, Cr, Co and Mn from soil and sewage sludge. J Sci Food Agric 36:794–798
Meiwes KJ (1995) Application of lime and wood ash to decrease acidification of forest soils. Water Air Soil Pollut 85:143–152
Melgar R, Benítez E, Nogales R (2009) Bioconversion of wastes from olive oil industries using the epigeic earthworm *Eisenia andrei*. J Environ Sci Health B 44:488–495
Merino I, Arevalo LF, Romero F (2005) Characterization and possible uses of ashes from wastewater treatment plants. Waste Manag 25:1046–1054
Mozaffari M, Rosen CJ, Russelle MP, Nater EA (2000) Chemical characterization of ash from gasification of alfalfa stem: implications for ash management. J Environ Qual 29:963–972
Muse JK, Mitchell CC (1995) Paper mill boiler-ash and lime byproducts as soil liming materials. Agron J 87:432–438
Naylor LM, Schmidt EJ (1989) Paper mill wood ash as a fertilizer and liming material: field trials. TAPPI J 72:199–206
Nogales R, Thompson R, Calmet A, Benitez E, Gómez M, Elvira C (1998) Feasibility of vermicomposting residues from oil production obtained using two stages centrifuge. J Environ Sci Health A 33:1491–1506
Nogales R, Melgar R, Benitez E (2006) Potential use of olive-waste ash from cogeneration plants as soil amendment. J Environ Sci Health B 41:1405–1415

Nurmesniemi H, Pöykiö R, Perämäki P, Kuokkanen T (2005) The use of a sequential leaching procedure for heavy metal fractionation in green liquor dregs from a causticizing process at a pulp mill. Chemosphere 61:1475–1484

Obernberger I, Supancic K (2009) Possibilities of ash utilisation from biomass combustion plants. In: Proceedings of the 17th European Biomass Conference & Exhibition, June/July 2009, Hamburg, 2009, pp 2373–2384

Obernberger I, Biedermann F, Widmann W, Riedl R (1997) Concentrations of inorganic elements in biomass fuels and recovery in the different ash fractions. Biomass Bioenergy 12:211–224

Ohno T, Erich MS (1993) Incubation-derived calcium carbonate equivalence of papermill boiler-ashes derived from sludge and wood sources. Environ Pollut 79:175–180

Omil B, Sanchez-Rodríguez F, Merino A (2011) Effects of ash applications on soil status, nutrition and growth of *Pinus radiata* D. Don plantations. In: Insam H, Knapp BA (eds) Recycling of biomass ashes. Springer, Heidelberg, pp 1–16

Pathan SM, Aylmore LAG, Colmer TD (2003) Properties of several fly ash materials in relation to use as soil amendments. J Environ Qual 32:687–693

Pitman RM (2006) Wood ash use in forestry – a review of the environmental impacts. Forestry 79:563–588

Sharma BM, Aggarwal RK, Kumar P (1990) Water retention and nutrient availability in a fly ash amended desert sandy soil: a study in vitro. Arid Soil Res Rehab 4:53–58

Someshwar AV (1996) Wood and combination wood-fired boiler ash characterisation. J Environ Qual 25:962–972

Ulery AL, Graham RC, Amrhein C (1993) Wood-ash composition and soil pH following intense burning. Soil Sci 156:358–364

Vance ED (1996) Land application of wood-fired and combination boiler ashes: an overview. J Environ Qual 25:937–944

Yeomans J, Bremner JM (1989) A rapid and precise method for routine determination of organic carbon in soil. Commun Soil Sci Plant Anal 19:1467–1476

Zhang FS, Yamasaki S, Nanzyo M (2002) Waste ashes for use in agricultural production: I. Liming effect, contents of plant nutrients and chemical characteristics of some metals. Sci Total Environ 284:215–225

Chapter 6
Effects of Ash Applications on Soil Status, Nutrition, and Growth of *Pinus radiata* D. Don Plantations

Beatriz Omil, Federico Sánchez-Rodríguez, and Agustin Merino

Abstract The aim of this study was to evaluate the effectiveness of multiple applications of biomass ash to acid soils. The study was carried out in two stands of *Pinus radiata* D. Don, aged 13 and 15 years, in the province of Lugo (northwest Spain). The soils in the stands were developed on lutites and migmatites. Experimental plots (each 1,225 m^2) were established, and the experimental treatments were as follows: control (untreated), ash (addition of 4.5 Mg dry matter ha^{-1} $year^{-1}$ in 2003, 2004, and 2005) and ash plus P (addition of ash plus phosphate fertilizer in 2003).

The ash was generated in a moving grate furnace, and had the following characteristics: pH 8.9–13.5, high concentrations of K, Ca, Mg, and P, and low N content and low concentration of heavy metals.

The responses of the forest stands, evaluated as the effects on forest nutrition and tree growth, were measured in 2005, 3 years after the initial treatment. The results showed that continuous fertilization with ash improved the nutritional status and growth of *Pinus radiata* D. Don stands, and resulted in increased contents of the main macronutrients in needles and soil.

6.1 Introduction

The Galician timber industry makes an important contribution to the regional economy. Forest land covers more than 60% of the total area, and the annual timber harvest is approximately 7,000,000 m^3. The wood is used in sawmills and to produce paper, particle board, and fiberboard. Lignocellulosic by-products generated in the latter industries are used in biomass incineration plants to meet increasing energy

B. Omil (✉), F. Sánchez-Rodríguez, and A. Merino
Forestry Faculty, Escuela Politécnica Superior, University of Santiago de Compostela, 27002 Lugo, Spain
e-mail: beatriz.omil@usc.es

H. Insam and B.A. Knapp (eds.), *Recycling of Biomass Ashes*,
DOI 10.1007/978-3-642-19354-5_6,

needs and to reduce waste production. The by-products mainly consist of bark (from pine, eucalyptus, and a small amount of birch, depending on the type of manufacturing) and to a lesser extent sand, dust, and panel fragments. The volume of these by-products created annually is approximately 900,000 m^3 and that of other by-products is approximately 40,000 m^3.

This source of energy is considered neutral from an environmental point of view, as it releases the same amount of CO_2 to the atmosphere as trees have removed. Combustion does not affect global warming or the greenhouse effect and has several advantages, such as a reduction in the use of fossil waste and the reuse of waste that has no value other than being a source of energy.

Biomass ash is produced as a result of this process (on average, combustion of wood produces 6–10% ash; Gaskin and Risse 2002). Ash is considered a non-hazardous waste (ERL codes 100101 and 100103, bottom ash and fly ash, respectively) and is therefore stocked at dumping sites. However, to reduce these stocks, alternative uses of the waste are being investigated, e.g., for production of enamel and glass (Xirokostas et al. 2001), as a building material for tracks or rural paths, for amation of coal mine (Gil-Bueno and Monterroso 1998; Seoane and Leirós 2001), as an absorbent for the removal of dichlorodiphenyldichloroethane (DDD) and dichlorodiphenyldichloroethylene (DDE) originating from pesticides (Gupta and Ali 2001), and as an additive in cement production (Van Der Sloot and Cnubben 2000). The latter use is not recommended because of the high level of C in the ashes. In this case, wood ash is usually used as a pozzolan (a siliceous and aluminous material). Despite not having binding properties, pozzolan reacts with finely divided calcium hydroxide in the presence of water to form compounds with cementing properties at room temperature (ASTM 1994).

Nevertheless, wood ash contains nutrients such as P, K, Ca, and Mg, which are present in relatively soluble forms (the NPK content is typically 0–1–5). Apart from these macronutrients, the waste contains oxides, hydroxides, and carbonates. The waste is therefore highly alkaline and contains low amounts of heavy metals (Erich and Ohno 1992; Korpilahti et al. 1998; Demeyer et al. 2001; Miller et al. 2002; Solla-Gullón et al. 2006). For all these reasons, application of wood ash to forest soils may be of interest as regards the environmental management of such waste, improvement of the nutritional status of forest plantations, and completing the CO_2 cycle (Torre-Minguela and Giraldo 2006). Fertilizer is added in an attempt to replenish the nutrients exported as a consequence of the extraction of biomass after final harvesting.

However, despite the high productivity of Galician forests, most *Pinus radiata* D. Don plantations are deficient in nutrients such as P, Mg, and Ca (Sánchez-Rodríguez et al. 2001; Zas 2003), which can be attributed to the strongly acidic soils and to the extraction of nutrients as a result of the management of the plantation in medium rotations (less than 40 years). Fertilization with wood ash would also contribute to the sustainability of the stands, intensive exploitation of which results in large losses of nutrients.

6.2 Materials and Methods

6.2.1 Site

This study was carried out in February 2003 in two *Pinus radiata* D. Don plantations, aged 13 and 15 years (Table 6.1), in the province of Lugo (northwest Spain) (43°13′ N, 7°39′ W and 43°07′ N, 7°49′ W respectively). Shrubs predominated in the understory vegetation (*Calluna vulgaris*, *Cytisus scoparius*, *Erica umbellata*, *Halimium lasianthum* subsp. *alyssoides*, *Ulex europaeus*, etc.).

The average annual precipitation during the study period (2003–2006) was 1,135 mm, and the average monthly minimum and maximum temperatures were 15.1 and 5.0°C, respectively.

6.2.2 Soil

The plots (Table 6.2) were established on different soils developed on migmatites and lutites both classified as Humic Umbrisol (FAO-UNESCO 1998) (Fig. 6.1). The soil in the first plot contained a high percentage of coarse particles (which create a sandy loam soil that favors good drainage), had a low pH (4.7), and had a C-to-N ratio of 19.3. In contrast, the soil in the second plot was of fine texture, with moderate permeability, was low hydraulic conductivity, had a high percentage of organic matter (C-to-N ratio 31.2), and had acid pH (4.8).

6.2.3 Experimental Design

The study comprised a random experimental design of a wood ash treatment program similar to fertilizer applications that are typically carried out in the study area. In 2003, four experimental plots of size 35 × 35 m^2 were established in quadruplicate. The experimental treatments were as follows: control (no treatment); WA [addition of 4.5 Mg dry matter (DM) ash ha^{-1}], and WAP (addition of 4.5 Mg DM ash ha^{-1} plus 0.1 Mg P_2O_5 ha^{-1}). In 2004 and 2005, the WA experimental plots were again treated with 4.5 Mg DM ash ha^{-1} (Fig. 6.2). It is important to

Table 6.1 Characteristics of the *Pinus radiata* D. Don stand plots

Location	Parent material	*t* (years)	SI (m)	*N* (trees ha^{-1})	*G* (m^2 ha^{-1})	Ho (m)	*v* (m^3 ha^{-1})	MAI (m^3 $ha^{-1}year^{-1}$)
Parga	Lutites	15	18.7	1465	18.5	15.3	113.4	7.6
Pino	Migmatites	13	23.1	792	15.5	15.7	104.8	8.1

t age of plantation, *SI* site index, *N* density, *G* basal area, *Ho* top height, *v* over bark volume, *MAI* mean annual increment

Table 6.2 Chemical analysis of the horizons in the soils developed on lutites and migmatites

	Color	pH (H_2O)	pH (KCl)	OM (%)	C/N (%)	S (%)	P (mg kg^{-1})	Ca (mg kg^{-1})	K (mg kg^{-1})	Mg (mg kg^{-1})	CECe ($cmol_c$ kg^{-1})	Cations (bases %)
Lutites												
Ah1	10YR3/3	4.8	4.1	10.3	31.2	0.08	3.3	38.4	42.8	14.8	21.3	6.4
Ah2	10YR3/4	4.7	4.2	6.9	18.2	0.04	4.6	24.4	29.6	10.6	7.6	9.8
Migmatites												
Ah1	10YR3/2	4.7	3.8	7.3	19.3	0.02	4.7	39.1	32.1	10.6	10.7	14.8
Ah2	10YR3/2	4.6	3.9	4.4	22.0	0.02	3.8	23.6	22.3	8.5	10.7	14.8

OM organic matter, *CECe* effective cation exchange capacity

Fig. 6.1 *Pinus radiata* D. Don plots: (**a**) lutites at Parga and (**b**) migmatites at Pino

emphasize that the phosphate fertilizer was only applied to the plots in 2003. The phosphate fertilizer (0–29–0) is partially disaggregated, and P release is therefore slower than from commercial fertilizers (e.g., superphosphate). This fertilizer is appropriate for acid soils and enables the amount of P applied to the soils to be monitored over time.

6.2.4 Fertilizer Composition

The chemical composition of biomass ash varies depending on the combustion technique used, the type of material used in the combustion (tree species and

Fig. 6.2 Appearance of the plots after application of wood ash

Table 6.3 Elements (kg ha^{-1}; Cd g ha^{-1}) applied in different treatments

	N	P	K	Ca	Mg	Mn	Fe	Zn	Cu	B	Cd	Ni
Control	–	–	–	–	–	–	–	–	–	–	–	–
WA	16.2	35.4	251.1	516.9	154.8	37.5	190.5	4.2	2.1	0.9	9.6	2.7
WAP	4.5	61.8	83.7	172.3	51.6	12.5	63.5	1.4	0.7	0.3	3.2	0.9

The nutrients applied in wood ash (*WA*) treatment are the sum of three applications of 4.5 Mg ha^{-1} in 2003, 2004, and 2005 (P 3 g kg^{-1}, K 19 g kg^{-1}, Ca 38 g kg^{-1}, Mg 11 g kg^{-1}, Mn 3 g kg^{-1}, Fe 14 g kg^{-1}, Zn 0.3 g kg^{-1}, Cu 0.2 g kg^{-1}, B 0.1 g kg^{-1}, Cd 0.7 mg kg^{-1}, Ni 0.2 g kg^{-1})

original material, bark, stem, braches, etc.), the storage conditions, etc. (Vance 1996; Blander 1997; Obernberger et al. 1997; Larsson and Westling 1998). The material used in this study was bottom ash generated in moving grate furnaces in three biomass plants in Spain (FINSA, Financiera Maderera). Chemically, this type of ash is less reactive than ash generated in fluidized bed combustors (equivalent neutralizing value 18% $CaCO_3$) since it contains a high percentage of unburned materials. The composition and the quantity of elements provided by such ash are shown in Table 6.3. The concentration of heavy metals is very low (bottom ash), especially in comparison with other waste such as sewage sludge and coal ash (below the limits established by the EU–European Community (1986) and the US Environmental Protection Agency).

However, there are some drawbacks associated with bottom ash which may hinder the spreading process, such as the high humidity (55%) and the high proportion of coarse elements (more than 20% larger than 4 mm) such as slug, unburned wood swath, and metals.

6.2.5 *Sampling and Chemical Analyses*

In each experimental plot, samples were collected from each soil layer, at a depth of 0–40 cm, from the center of each plot, following a zigzag route. Soil samples were mixed to ensure homogeneity, dried at 40°C, and sieved through a 2-mm screen; microoelements and macroelements were extracted by the Mehlich 3 procedure (Mehlich and Mehlich 1984). Soil pH was measured in H_2O and KCl (0.1 M). For elemental analysis (CNS), the soil samples were ground in a mortar to obtain a fine powder.

Sampling was carried out in spring and autumn. Needle samples were also collected, approximately 25% from the floor of the plot (Ballard and Carter 1986), approximately 25% from the upper third, approximately 25% from the growth of the year, and approximately 25% from the sunniest branches (Will 1985). The needle samples were dried at 60°C to constant weight and were ground for CNS analysis by inductively coupled plasma optical emission spectrometry.

6.2.6 *Statistical Analysis*

Analysis of variance was used to test for effects of multiple wood ash application on soil chemical properties and foliar concentration of macroelements and microelements, with the PROC GLM procedure of SAS (SAS Institute 2004).

Normal diameter and total height were measured annually in all trees within the plots at the end of the growing season for a period of 3 years. The breast diameter was measured in two directions with a caliper, to an accuracy of less than 1 mm. Tree height was measurement with a Vertex III hypsometer. A volume equation based on the allometric model of Schumacher and Hall was used to calculate the tree volumes. The parameters of this model were estimated by Castedo (2004) by use of the following expression:

$$v = 0.000048 \times d^{2.0062} \times \mathrm{ht}^{0.86691},$$

where v is the bark volume (m^3) of individual trees; d is the breast diameter in centimeters, and ht is the total height in meters. The total height and breast diameters were compared by repeated measurement analysis with the PROC GLM procedure of SAS (SAS Institute 2004) after prior confirmation of the assumptions of equal (Levene test), normal (Kolmogorov–Smirnov test), and independent variance. The mathematical model used was as follows:

$$yij = \mu + \beta xit + Ti + Dj + Ti \times Dj + \varepsilon ij,$$

where y_{ij} is the random variable representing the value in the *j*th observation of the *i*th treatment, μ is a constant representing the mean response of the variable; βxit models the linear relationship between the response and the covariate (initial height or diameter measured before the application of ashes), T_i and D_j are the effects of treatment *i* (control, WA, and WAP) and the time *j* (0,1,2,...4 measurements), respectively, $T_i \times D_j$ is the interaction effect of treatment *i* by time *j*, and ε_{ij} is the experimental error.

6.3 Results

6.3.1 Soil

Soil pH and Ca, Mg, K, P, and S levels increased slightly after ash application. No leaching loss was observed after several applications of ash. Furthermore, microbial activity increased and ammonium and nitrate concentrations decreased as a consequence of N immobilization. No effects due to heavy metals were observed in the soil solution, and the availability of Mn and Zn increased slightly after the third application of ash.

6.3.2 Nutrition

The *Pinus radiata* plantations were initially deficient in P, K, and Mg (Table 6.4), as often found in this type of acid soil plantation (Mesanza et al. 1993; Romanyá and

Table 6.4 Concentrations of nutrient elements in the needles (mg g^{-1})

	N	P	K	Ca	Mg	N/P
Lutites	13.4	0.9	4.3	1.3	1.0	18.0
Migmatites	15.0	0.9	5.6	1.4	0.8	17.7
Will (1985) and Lambert (1984)[a]	15	>1.4	>5.0	>1.0	>1.0	–
Zas et al. (2002)[b]	16.9	1.1	6.2	2.1	0.8	15.4
Sánchez-Rodríguez et al. (2002)[c]	16.4	0.8	7.5	1.7	0.7	20.5

WAP wood ash plus P_2O_2

[a]Optimal values for *Pinus radiata* D. Don plantations

[b]Foliar concentrations in Galicia for *Pinus radiata* young plantations

[c]Average concentrations of nutrient elements in the needles of *Pinus radiata* plantations on different soil parent material

Vallejo 1996). In contrast, foliar N concentrations were sufficient (15 mg g^{-1}). Repeated application of wood ash to the soil did not have any effect on foliar concentration 3 years after the initial treatment. Although the concentrations of basic cations such as Ca^{2+} and Mg^{2+} increased significantly in the soil, foliar analysis did not show any significant changes in the needles after the treatment. This may be the result of tree growth and the consequent dilution effect.

Supplemental fertilization with slow release of phosphorus (WAP treatment) increased the foliar concentration of the element throughout the study (Fig. 6.3). The differences between treatments are more significant for the soils over migmatites than for those over lutites. However, repeated application of wood ash did not increase the levels of P in needles.

6.3.3 *Heavy Metals*

Repeated application of ash did not produce significant changes in foliar concentrations of any of the elements. However, the application produced an increase in the concentration of Zn, Cu, and Ni in needles in the plots on lutites (Table 6.5). This was not observed in 2006. The foliar concentration of Cd was closely related to soil acidity and therefore the concentration of Cd decreased from the third application of ash onwards.

6.3.4 *Tree Growth*

Plots on lutites and migmatites were selected for this study in order to compare two plantations of the same age, but located over two different geological materials.

The following figures include the estimated marginal means of the dasymetric variables: total height and normal diameter (corrected in the covariance function, data from the plots before the treatment in 2003 are used as a covariate). A statistical analysis comparing between levels of the factor "treatment" corresponding to each level of the factor "time" and the levels of the repeated means is also included.

Different effects were observed in both types of plots. In the first plot (Fig. 6.4), the statistical analysis revealed significant increases in height and diameter growth as a result of the application of ash and phosphorus, mainly from 2005 onwards. The time–treatment interaction revealed some differences between the treatments, with benefits to both plots. In the second plot (Fig. 6.5) increases in the normal diameter for WA and WAP treatments were significant. The different response of the plantations may be due to the differences in the nutritional state, as indicated by the different site index values (18.7 and 23.1 m in the lutites and migmatites, respectively).

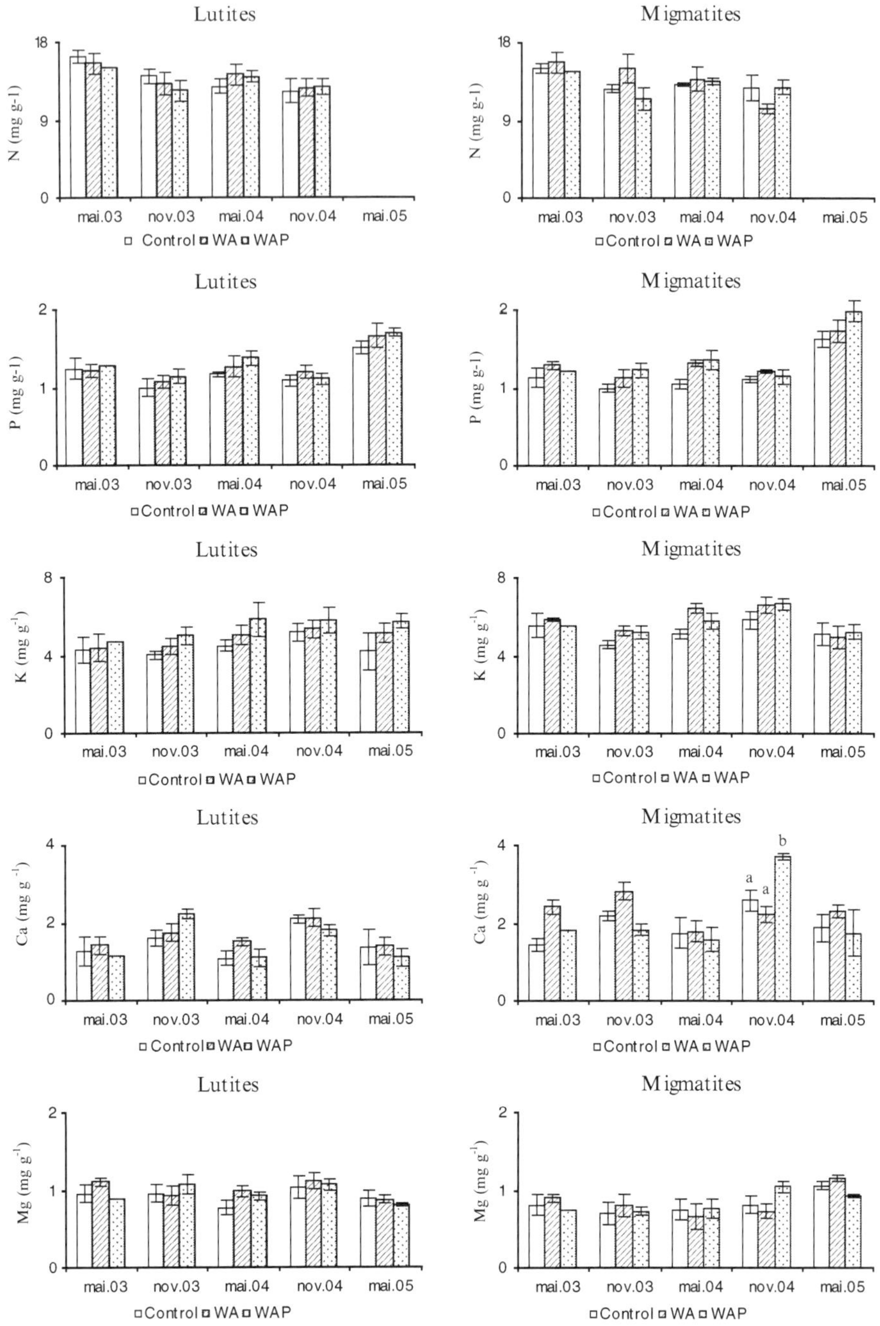

Fig. 6.3 Foliar macronutrients in soils over lutites and migmatites. *Control* untreated, *WA* applications of 4.5 Mg wood ash ha^{-1} for three consecutive years (2003, 2004, 2005), *WAP* single applications of 4.5 Mg wood ash ha^{-1} and 0.1 Mg P_2O_5 ha^{-1}

Table 6.5 Average heavy metal concentrations in *Pinus radiata* D. Don needles (Mg kg^{-1} dry matter)

Nutrient	Treatment	May 03		Nov 03		May 04		Nov 04		May 05		Nov 05	
		Mean	SD	Mean	SD	Mean	SD	Mean	SD	Mean	SD	Mean	SD
Zn	Control	24.6	3.5	27.3	2.2	28.0	4.2	22.5	1.6	25.6	2.1	19.3	1.0
	WA	18.7	2.4	29.4	2.9	25.4	4.9	27.6	1.9	30.1	2.0	24.9	1.7
	WAP			26.1	2.4	24.6	2.3	27.2	0.7	28.4	2.4	28.7	3.8
Cu	Control	3.34	0.85	2.90	0.22	3.27	0.22	3.69	0.13	3.80	0.22	3.33	0.14
	WA	3.34	0.85	2.90	0.22	3.27	0.22	3.69	0.13	3.80	0.22	3.33	0.14
	WAP	0.01	0.01	3.11	0.60	3.55	0.15	4.41	0.15	3.89	0.16	3.56	0.21
Cd	Control	0.03	0.01	0.05	0.01	0.12	0.02	0.24	0.02	0.12	0.01	0.06	0.01
	WA	0.03	0.01	0.11	0.04	0.11	0.03	0.29	0.03	0.21	0.12	0.05	0.01
	WAP			0.04	0.01	0.16	0.02	0.29	0.02	0.14	0.05	0.05	0.01
Ni	Control	1.40	0.30	1.23	0.24	1.62	0.16	1.34	0.13	0.67	0.23	1.19	0.10
	WA	1.35	0.29	1.24	0.28	1.62	0.24	1.37	0.11	0.51	0.06	1.45	0.24
	WAP			1.05	0.10	1.77	0.18	1.68	0.21	0.45	0.04	1.05	0.00
Zn	Control	35.1	1.1	39.6	3.1	26.3	0.3	26.6	0.9	31.1	1.8	35.5	2.8
	WA	33.1	1.1	39.0	3.0	24.8	2.2	30.5	0.4	23.9	0.4	43.1	1.4
	WAP			29.2	0.9	25.7	2.4	36.2	0.7	26.1	2.0	31.3	2.0
Cu	Control	4.89	0.36	3.50	0.10	3.44	0.05	3.92	0.23	3.72	0.31	3.88	0.11
	WA	4.63	0.45	3.87	0.22	3.72	0.28	3.67	0.15	4.19	0.45	3.69	0.13
	WAP			3.44	0.51	3.42	0.33	3.70	0.06	3.37	0.09	3.73	0.15
Cd	Control	0.09	0.02	0.03	0.02	0.10	0.01	0.09	0.01	0.10	0.02	0.04	0.01
	WA	0.06	0.01	0.14	0.02	0.18	0.03	0.29	0.10	0.17	0.01	0.01	0.00
	WAP	0.00	0.00	0.11	0.03	0.10	0.01	0.19	0.02	0.16	0.04	0.05	0.01
Ni	Control	1.14	0.18	1.48	0.22	1.27	0.15	1.16	0.04	0.55	0.05	1.11	0.02
	WA	1.83	0.24	1.09	0.10	1.32	0.16	1.27	0.00	1.22	0.21	1.10	0.02
	WAP			1.37	0.15	1.27	0.04	1.01	0.09	0.71	0.07	1.12	0.02

SD standard deviation

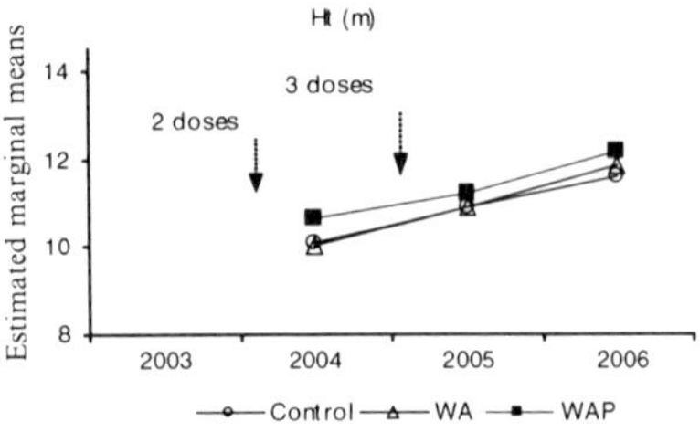

A: Comparison of means between levels of factor "treatment" for each level of factor "time"					B: Repeated measures analysis[a]		
Treat	Time t_0	t_1	t_2	t_3		*F*	*P*
Control		a	a	a	Covariate	31641.9	**0.000**
WA		b	b	b	Treat	42.5	**0.000**
WAP		a	b	c	Time	324.4	**0.000**
					Treat*Time	39.2	**0.000**

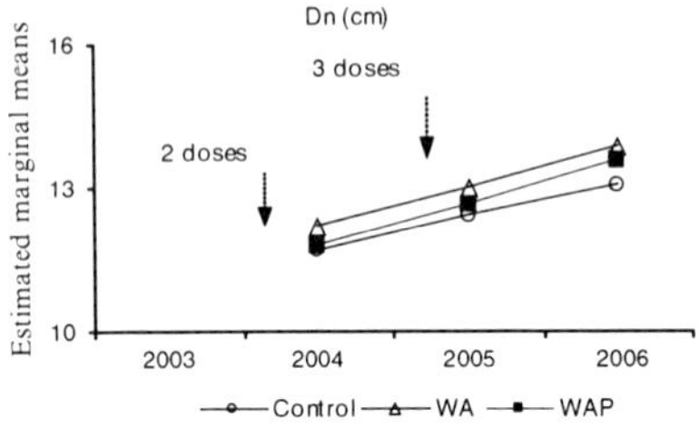

A: Comparison of means between levels of factor "treatment" for each level of factor "time"					B: Repeated measures analysis[a]		
Treat	Time t_0	t_1	t_2	t_3		*F*	*P*
Control		a	a	a	Covariate	27037.6	0.000
WA		b	b	b	Treat	32.4	0.000
WAP		a	a	a	Time	23.4	0.000
					Treat*Time	7.6	0.000

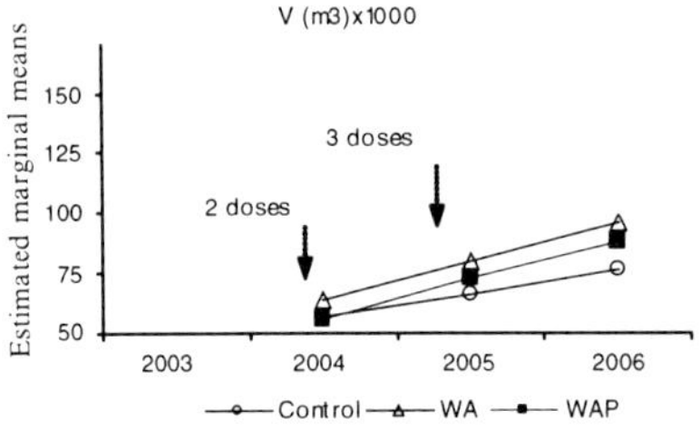

A: Comparison of means between levels of factor "treatment" for each level of factor "time"					B: Repeated measures analysis[a]		
Treat	Time t_0	t_1	t_2	t_3		*F*	*P*
Control		a	a	a	Covariate	18486.4	0.000
WA		b	b	b	Treat	34.7	0.000
WAP		b	b	b	Time	60.2	0.000
					Treat*Time	27.8	0.000

Fig. 6.4 Estimated marginal means of *Pinus radiata* D. Don in plots over lutites. *Control* untreated, *WA* application of 4.5 Mg wood ash ha^{-1} for three consecutive years (2003, 2004, 2005), *WAP* single applications of 4.5 Mg wood ash ha^{-1} and 0.1 Mg P_2O_5 ha^{-1}. In the repeated measures analysis, the sphericity assumption was not valid and therefore values of P-Huynh Feldt were used in univariate tests of their interactions. *Ht* total height, *Dn* normal diameter

6.4 Discussion

The bottom ash used in this study was generated in moving grate furnaces and did not contain fly ash. This is a major difference from fluidized bed combustors, which produce an ash mix. The analysis thus revealed lower concentrations of N, P, K, Mg, and S than those reported by other authors (Ohno and Erich 1990; Korpilahti et al. 1998; Demeyer et al. 2001; Miller et al. 2002; Solla-Gullón et al. 2006). In accordance, heavy metal concentrations were relatively low compared with those of other waste products used in agriculture, e.g., slurry from wastewater treatment plants and coal ash (Arvidsson and Lundkvist 2002; Hytönen 2003).

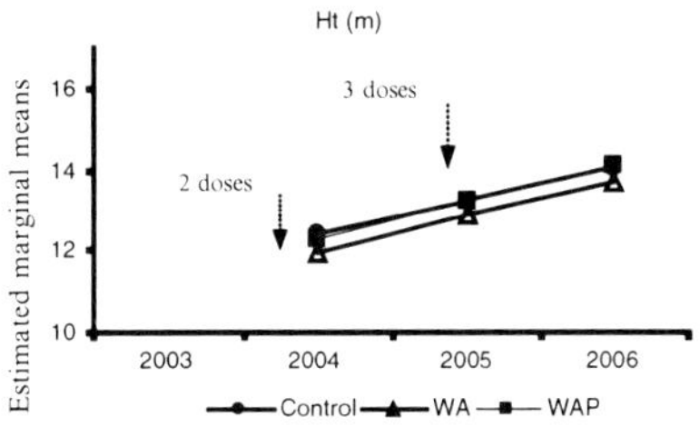

A: Comparison of means between the levels of factor "treatment" for each level of factor "time"					B: Repeated measures analysis[a]		
Treat	Time					*F*	*P*
	t_0	t_1	t_2	t_3	Covariate	30700	0.000
Control		a	b	b	Treat	29.99	0.000
WA		a	a	a	Time	428.38	0.000
WAP		b	b	b	Treat*Time	2.44	0.111

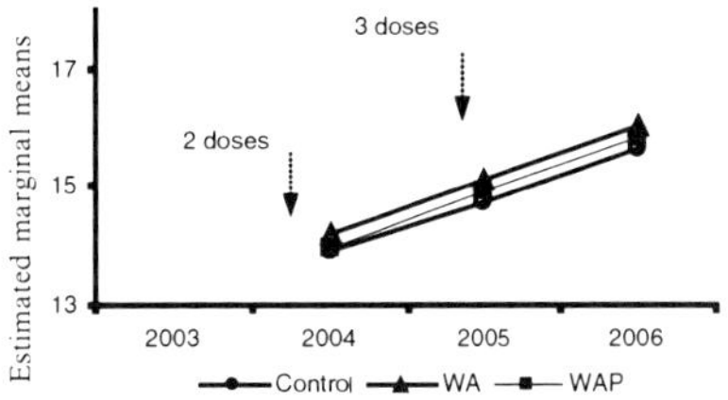

A: Comparison of means between the levels of factor "treatment" for each level of factor "time"					B: Repeated measures analysis[a]		
Treat	Tiemp o					*F*	*P*
	t_0	t_1	t_2	t_3	Covariate	4260.1	0.000
Control		b	a	a	Treat	3.7	0.024
WA		b	a	a	Time	4.8	0.027
WAP		a	a	a	Treat*Time	1.1	0.325

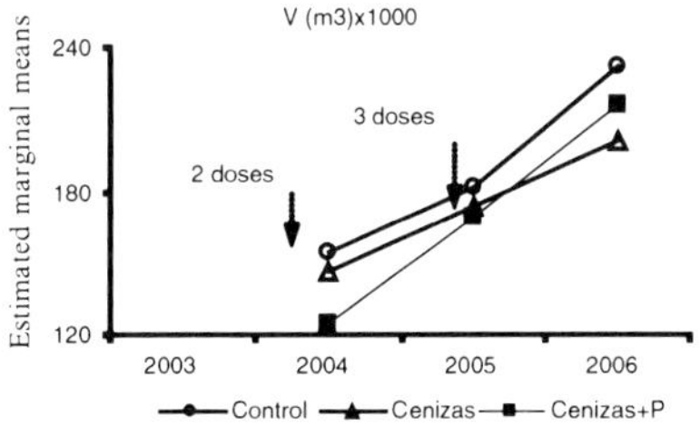

A: Comparison of means between the levels of factor "treatment" for each level of factor "time"					B: Repeated measures analysis[a]		
Treat	Tiemp o					*F*	*P*
	t_0	t_1	t_2	t_3	Covariate	1050.2	0.000
Control		a	a	a	Treat	2.6	0.070
WA		a	b	b	Time	2.1	0.147
WAP		a	a	a	Treat*Time	0.6	0.540

Fig. 6.5 This figure is a result of *Pinus radiata* D. Don for plots over migmatites. *Control* untreated, *WA* application of 4.5 Mg wood ash ha^{-1} for three consecutive years (2003, 2004, 2005), *WAP* single applications of 4.5 Mg wood ash ha^{-1} and 0.1 Mg P_2O_5 ha^{-1}. In the repeated measures analysis, the sphericity assumption was not valid and therefore values of P-Huynh Feldt were used in univariate tests of their interactions

6.4.1 Soil

The soils under study are acid, as are most forest soils in the region, mainly due to the instability of the parent rock and to the strong leaching of the base cations. Also, the plant growth itself generates soil acidity. This implies a great obstacle to the growth of forest species since it reduces the availability of essential nutrients such as P (Sánchez-Rodríguez et al. 2002; Merino et al. 2005).

In this study, the application of ash slightly and temporarily increased the pH of the solid soil fraction, resulting in higher availability of elements such as K, Ca, and Mg. The availability was proportional to the quantity of ash applied. The concentration of Ca increased greatly after the second application. Thus, a direct relationship between the application and the availability of the element was observed.

Nevertheless, the increases were less than those reported for soil treated with fly ash (Ohno and Erich 1990; Saarsalmi et al. 2001; Hytönen 2003; Solla-Gullón et al. 2006).

The behavior of P was different, and a low response to the different treatments was observed. This may be due to the low availability of P in the ash, as well as to the chemical and biological reactions that this element undergoes in the soil. In acid soils, phosphate ions tend to precipitate with Fe and Al, forming insoluble compounds and thus reducing the concentrations of $H_2PO_4^-$ and H_2PO_4 taken up by plants (Garcia-Rodeja and Gil-Sostres 1997). It is also possible that the availability of P decreases owing to microbial immobilization, a process that is accelerated by the application of organic waste to the soil (Salas et al. 2003).

6.4.2 Nutrition

Results of foliar analyses may differ depending on environmental factors. The date of sampling, the age of the plants, and the type of parent material found below the plantation must be taken into account.

The foliar concentration of N tends to decrease in response to application of biomass ash, as expected from the low N content in the ash and its immobilization in the soil. However, an increase in N concentration has been observed in some peat soils, which can be attributed to a higher mineralization as a consequence of an increase in pH and availability of nutrients (Weber et al. 1985). In a prior study, Solla-Gullón et al. (2006) also detected an increase in foliar N concentration in *Pseudotsuga menziesii* plantations in Galicia.

In this study, the foliar values of P were prone to increase, as observed by Moilanen et al. (2002), Ludwig et al. (2002), and Solla-Gullón et al. (2008). This positive response may be the result of the symbiotic association between ectomycorrhizal fungi that colonize the ash. These fungi increase the solubility of the content of P in the ashes, thereby promoting the uptake of ash by trees (Mahmmood et al. 2003). Use of a diagnostic system based on the N-to-P ratio showed that a balance between N and P was achieved, since it did not exceed the normal ratio ranging from 6 to 16 (Raupach 1967). These values are similar to the average value of 12.0 reported by Solla-Gullón et al. (2008) for *Pinus radiata* in Galicia.

Foliar K concentration also increased slightly in the WA and WAP plots, confirming the relationship between plant and soil concentration (Zas 2003). Increases in plant K concentration in response to the application of ash have been reported earlier (Moilanen et al. 2002; Solla-Gullón et al. 2008). However, no differences were observed by Hytönen (2003).

Despite the higher availability of Ca and Mg in the soil, foliar tree analysis did not reveal significant increases in concentration for these elements. This is consistent with the findings of some studies in which concentrations did not increase, or increased only slightly (Hytönen 2003; Moilanen et al. 2002; Ludwig et al. 2002). Other studies in which greater amounts of ash were applied revealed increases in the foliar concentration, which lasted for a few years (Solla-Gullón et al. 2008; Arvidsson and Lundkvist 2002).

6.4.3 Heavy Metals

Soil concentrations of Cu, Ni, Zn, and Cd were similar to the average values reported by Merino et al. (2005) for forest soils. Repeated application of ash did not increase the level of heavy metals in the soil, and the levels of extractable Ni decreased owing to the slight increase in pH, which reduced the solubility of Ni. The concentrations of Cd did not change in either the soil or the tree needles. This is consistent with the results reported by other authors such as Helmisaari et al. (2009).

6.4.4 Tree Growth

Earlier studies revealed that *Pinus radiata* shows a good response to fertilization (Sánchez-Rodríguez et al. 2002; Omil et al. 2005; Solla-Gullón et al. 2008). However, this response varies depending on the age of the trees, on the density, and on the nutritional needs (Zas 2003). The demand for elements such as N and P increases during the first years of life, a maximum of 6–8 years (Turner and Lamber 1986). Ca and Mg, which do not suffer retranslocation before needle fall (Ericsson 1994), may cause problems in forest crops since the demand for these elements by higher plants tends to increase with age (Turner and Lamber 1986).

Analysis of covariance revealed that improvement in soil nutritional status leads to higher tree growth from the second year of treatment onward, confirmingthe findings of other studies (Bonneau 1995; Vesterinen 2003). This is due to the site quality (ecological conditions, soil physical properties, weather, and water availability of the trees): under adverse conditions fertilization does not improve soil nutritional status. Thus, the response was more significant in plots on lutites (of lower site quality) than in plots on migmatites. This confirms the results reported by other authors, mainly for rich soils (Silfverberg and Huikari 1985; Ferm et al. 1992; Emilsson 2006). Production may be improved by application of a higher dose of ash, although there is a risk of leaching contamination and increased heavy metal concentrations.

The application of other types of waste such as slag and dairy sludge was found to increase tree height growth (Virgel-Mentxaka 2002; Omil et al. 2005). The combination of ashes with certain organic wastes has considerably increased the growth of agricultural crops (see Chap. 4, Nieminen 2011). However, the amount of N applied must be taken into account. Prior studies revealed that the tree structure of this species may be affected by N and P stresses (Will 1985). Although height growth is not usually altered, radial growth in both stem and branches was most sensitive to addition of these elements. These changes have important implications such as the loss of apical dominancy in fertilized trees and possible stem distortions, resulting in a decrease in the economic value (Will 1985; Hopmans and Chappell 1994).

6.5 Conclusions

Fertilization with bark ash improved the nutritional status of *Pinus radiata* plantations, and increased the contents of the main macronutrients in the needles. However, some limiting nutrients such as P did not exceed critically low levels, so there is some room for improvement in the fertilization treatments. The effects on the needles were also inconsistent, and delay and the intensity of the effects depended on the time of application. The third application of ash significantly improved the diameter and height (and so the volume) growth in one of the plots (lutites) relative to the control treatment. It may be concluded that only after careful site evaluation ash application is indicated, otherwise the costs may exceed the benefit.

References

Arvidsson H, Lundkvist H (2002) Needle chemistry in young Norway spruce stands after application of crushed wood ash. Plant Soil 238:159–174

ASTM (1994) Standard specification for coal fly ash and raw or calcined natural pozzolan for use in concrete. ASTM C618-94. American Society for Testing and Materials), West Conshohocken

Ballard TM, Carter RE (1986) Evaluating forest stand nutrient status. Land management report no. 20. Ministry of Forests, Victoria

Birk EM (1994) Fertilizer use in the management of pine and eucalyptus plantations in Australia: a review of past and current practices. State Forest of New South Wales N Z J For Sci 24:289–320

Blander B (1997) The inorganic chemistry of the combustion of aspen wood with added sulphur. Biomass Bioenergy 12:289–293

Bonneau M (1995) Fertilisation des Forêts dans les Pays Tempérés. École Nationale du Génie Rural des Eaux et des Forêts (ENGREF), Nancy

Castedo F (2004) Modelo dinámico de crecimiento para las masas de *Pinus radiata* D. Don en Galicia. Simulación de alternativas selvícolas con inclusión de riesgo de incendio. Doctoral thesis, Escuela Politécnica Superior, Universidad de Santiago de Compostela, Santiago de Compostela

Demeyer A, Voundinkana JC, Verloo MG (2001) Characteristics of wood ash and influence on soil properties and nutrient uptake: an overview. Bioresour Technol 77:287–295

Emilsson S (2006) International handbook from extraction of forest fuels to ash recycling. Skogsstyrelsen/Swedish Forest Agency, Karlstad

Ericsson T (1994) Nutrient dynamics and requirements of forest crops. N Z J For Sci 24:133–168

Erich MS, Ohno T (1992) Phosphorous availability to corn from wood ash amended soils. Water Air and Soil Pollution 64:475–486

European Communities (1986) Council directive on the protection environment and in particular on the soil when sewage sludge is used in agriculture (86/278 EEC). Off J Eur Communities 181:6–12

FAO-UNESCO (1998) Clasificación de Suelos FAO. Base de Referencia de los Suelos del Mundo. FAO, Rome

Ferm A, Hokkanen A, Moilanen M, Issakainen J (1992) Effects of wood bark ash on the ground and nutrition of a Scots pine afforestation in central Finland. Plant Soil 147:305–316

Garcia-Rodeja I, Gil-Sostres F (1997) Prediction of parameters describing phosphorous desorption kinetics in soils of Galicia (northwest Spain). J Environ Qual 26:1363–1369

Gaskin J, Risse M (2002) Best management practices for wood ash agricultural soil amendment. Bulletin 1142. Cooperative Extension Service, College of Agricultural and Environmental Sciences, The University of Georgia, Athens, pp 1–4

Gil-Bueno A, Monterroso C (1998) Metodoloxía de restauración de escombreiras na minería de carbón. In: II Congreso de Enxeñería da Paisaxe. Asociación Española de Ingeniería del Paisaje, Guipúzcoa, pp 101–109

Gupta VK, Ali I (2001) Removal of DDD and DDE from wastewater using bagasse fly ash a sugar industry. Water Res 35:33–40

Helmisaari HS, Saarsalmi A, Kukkola M (2009) Effects of wood ash and nitrogen fertilization on fine root biomass and soil foliage nutrients in a Norway spruce stand in Finland. Plant Soil 314:121–132

Hopmans P, Chappell HN (1994) Growth response of young, thinned Douglas-fir stands to nitrogen fertilizer in relation to soil properties and tree nutrition. Can J For Res 24:1684–1688

Hytönen J (2003) Effects of wood peat and coal ash fertilization on Scots pine foliar nutrient concentrations and growth on afforested former agricultural peat soils. Silva Fenn 37:219–234

Korpilahti A, Moilanen M, Finér L (1998) Biomass ash utilization in Finland. In: International biomass ash workshop, Graz, Austria, Oct 1998

Lambert MJ (1984) The use of foliar analysis in fertilizer reseach. In: Proceeding of the IUFRO Symposium on Site and Productivity of Fast Growing Plantations, Pretoria, South Africa. 269–291

Larsson PE, Westling O (1998) Leaching of wood ash and lime products: laboratory study. Scand J For Res 2:17–22

Ludwig B, Rumpf S, Mindrup M, Meiwes KJ, Khanna P (2002) Effects of lime and wood ash on soil-solution chemistry, soil chemistry and nutritional status of a pine stand in northern Germany. Scandinavian Journal of Forest Research 17:225–237

Mahmmood S, Finlay RD, Framson AM, Wallander H (2003) Effects of hardened wood ash on microbial activity. Plant growth and nutrient uptake by ectomycorrhizal spruce seedlings. FEMS Microbiol Ecol 43:121–131

Mehlich A, Mehlich N (1984) Extractant: a modification of Mehlich no. 2 extractant. Commun Soil Sci Plant Anal 15:1409–1416

Merino A, Balbóa MA, Ródríguez-Soalleiro R, Álvarez-González JG (2005) Nutrient exports under different harvesting regimes in fast-growing forest plantations in southern Europe. For Ecol Manage 207:325–339

Mesanza JM, Casado H, Castillo FJ (1993) Nutrient concentrations in *Pinus radiata* D. Don needles in the Basque Country (Spain): a preliminary classification of parameters and sites. Ann For Sci 12:85–98

Miller BB, Dugwell DR, Kandiyoti R (2002) Partitioning of trace elements during the combustion of coal and biomass in a suspension-firing reactor. Fuel 81:159–171

Moilanen M, Silverberg K, Hokannen TJ (2002) Effects of wood-ash on the tree growth vegetation and substrate quality of a drained mire: a case study. For Ecol Manage 171:321–338

Naylor LM, Schmidt EJ (1986) Agricultural use of wood ash as a fertilizer and liming material. TAPPI J 69:114–119

Nieminen JK (2011) Wood ash effects on soil fauna and interactions with carbohydrate supply: a mini-review. In: Insam H, Knapp BA (eds) Recycling of biomass ashes. Springer, Heidelberg, pp 46–56

Obernberger I, Biedermann F, Widmann W, Riedl R (1997) Concentrations of inorganic elements in biomass fuels and recovery in the different ash fractions. Biomass Bioenergy 12:211–224

Ohno T, Erich MS (1990) Effect of wood ash application on soil pH and soil test nutrient level. Agric Ecosyst Environ 32:433–438

Omil B, Solla-Gullón F, Merino A (2005) Respuesta en crecimiento de una plantación de *Pinus radiata* D. Don. a la aplicación de lodos de lechería estabilizados con cal. Cuad Soc Esp Cienc For 5:191–196

Raupach M (1967) Soil and fertilizer requirements for forests of *Pinus radiata*. Adv Agron 19:307–353

Romanyá J, Vallejo R (1996) Nutritional status and deficiency diagnosis of *Pinus radiata* plantations in Spain. For Sci 42:192–197

Saarsalmi A, Mälkönen E, Pirainen S (2001) Effects of wood ash fertilization on forest soil chemical properties. Silva Fenn 35:355–368

Salas AM, Elliot ET, Westfall DG, Cole CV, Six J (2003) The role of particulate organic matter in phosphorous cycling. Soil Sci Soc Am J 67:181–189

Sánchez-Rodríguez F (2001) Estudio de la Calidad de Estación, Producción y Selvicultura de Pinus radiata D. Don en Galicia. Doctoral Thesis. Santiago de Compostela University

Sánchez-Rodríguez F, Rodríguez-Soalleiro R, Español E, López CA, Merino A (2002) Influence of edaphic factors and tree nutritive status on the productivity of *Pinus radiata* D. Don plantations in northwestern Spain. For Ecol Manage 171:181–189

SAS Institute (2004) SAS STAT® user's guide. SAS Institute, Cary

Seoane S, Leirós MC (2001) Acidification–neutralization processes in a lignite mine spoil attended with fly ash or limestone. J Environ Qual 30:1420–1431

Silfverberg K, Huikari O (1985) Wood-ash fertilization on drained petlands. Folia For 633:1–25

Solla-Gullón F, Santalla-Seoane M, Rodríguez-Soalleiro R, Merino A (2006) Nutritional status and growth of a young *Pseudotsuga menziesii* plantation in a temperate region after application of wood-bark ash. For Ecol Manage 237:312–321

Solla-Gullón F, Santalla M, Pérez-Cruzado C, Merino A, Rodríguez-Soalleiro R (2008) Response of *Pinus radiata* seedlings to application of mixed wood-bark ash at planning in a temperate region: nutrition and growth. For Ecol Manage 255:3873–3884

Torre-Minguela C, Giraldo GA (2006) Instalación térmica con biomasa residual en una industria del sector maderero. Proyecto fin de carrera. Escuela Técnica Superior de Ingenieros Industriales, Universidad de Valladolid, Valladolid

Turner J, Lamber MJ (1986) Nutrition and nutritional relationships of *Pinus radiata*. Annu Rev Ecol Syst 17:325–350

Van Der Sloot HA, Cnubben PAJP (2000) Verkennende evaluatie kwaliteitsbeinvloeding poederkool pliegas-bijtoken van biomassa in een poederkoolcentrale of bijmengen van biomassa-assen met poederkoolviliegas. ECN-C-00-058

Vance ED (1996) Land application of wood fired and combination boiler ashes: an overview. J Environ Qual 25:937–944

Vesterinen P (2003) Wood ash recycling state of the art in Finland and Sweden. VTT Processes, Jyväskylä

Virgel-Mentxaka S (2002) Efecto de la aplicación de escoria siderúrgica LD y fertilizante NPK en sistemas agrarios. Doctoral thesis, Universidad del País Vasco, Leioa

Weber A, Karsisto M, Leppanen R, Sudman V, Skujins J (1985) Microbial activities in a histosol. Effects of wood ash and NPK fertilizers. Soil Biol Biochem 17:291–296

Will GM (1985) Nutrient deficiencies end fertiliser user in New Zealand exotic forests. FRI bulletin no. 97. FRI, Rotorua

Xirokostas N, Moutsatsou A, Arvelakis S, Koukios E (2001) Study of the behaviour of agricultural ashes concerning the combustion process of the raw material. University of Antennas

Zas R (2003) Interpretación de las concentraciones foliares en nutrientes en plantaciones jóvenes de *Pinus radiata* D. Don en tierras agrarias en Galicia. Invest Agrar Sist Recur For 12:3–11

Zas R, Serrada R (2002) Foliar nutrient status and nutritional relationships of young Pinus radiata D. Don plantations in northwest Spain. Forest Ecology and Management 174:167–176

Chapter 7
Possible Use of Wood Ash and Compost for Improving Acid Tropical Soils

Blaise Pascal Bougnom, Brigitte Amalia Knapp, François-Xavier Etoa, and Heribert Insam

Abstract Infertility of acid soils is a major limitation to crop production on highly weathered and leached soils throughout the world. The main characteristics of these soils are their low pH, low levels of organic matter, Ca, Mg, P, or Mo deficiency, Al or Mn toxicity, or both, and very low mineralization and nitrification rates. Lime is generally recommended to correct soil acidity, but lime is unaffordable for resource-poor farmers in the tropics. Many alternatives have been proposed, and among them products from organic waste materials, e.g., composts, have proven to be an efficient alternative to the use of lime. Wood ash is a potential source of trace elements, nutrients, and lime. Wood ash could be used as an additive to fertilizer, and wood ash admixture to organic wastes prior to composting is known to improve compost quality and may reduce the amount of compost required to raise the pH to suitable levels. Wood ash compost as a liming agent as a replacement for lime could potentially aid in remediating acidity and base deficiency as well as boosting the soil microbial pool in tropical agricultural soils.

7.1 Introduction

Agricultural primary production is essential for maintaining human life. Sustaining the productivity of soils is important for future generations, but the way to maintain productivity is often disputed. Intensive agriculture is based on the use of large quantities of pesticides and other chemical substances aiming at increasing yields, but the price to pay is the deterioration of soil quality and the environment, water pollution, the emergence of new pathogens that are more and more resistant to pesticides, and the threat to human health from the consumption of pesticide residues

B.P. Bougnom (✉) and F.-X. Etoa
Laboratory of Microbiology, Department of Biochemistry, University of Yaoundé I, P.O. Box 812, Yaoundé, Cameroon
e-mail: bpbougnom@uy1.uninet.cm

B.A. Knapp and H. Insam
Institute of Microbiology, University of Innsbruck, Technikerstraße 25 d, 6020 Innsbruck, Austria

H. Insam and B.A. Knapp (eds.), *Recycling of Biomass Ashes*,
DOI 10.1007/978-3-642-19354-5_7,

entering the food chain as well as inhalation of toxic gases. These practices, however, are beyond the reach of resource-poor farmers in the tropics, because of their high cost. On the other hand, agricultural and forestry products are exported from tropical to western countries; together with nutrients like nitrogen, phosphorus and potassium, present in this biomass. Sustainability issues are thus becoming more important, reflecting the need for long-term fertility and environmental protection. A sustained agricultural system is one in which the sum of income extracted every year is sustained over years without altering the natural resource levels (Yunlong and Smit 1994). Organic farming principles and objectives are achieving good crop yields by using techniques which minimize the human impact on the environment (Rigby and Caceres 2001). Organic farming can allow resolution of the problem of disposal of organic wastes that human beings have to get rid of and which are a valuable source of nutrients for plants, and serve the purpose of soil conditioners. That allows the carbon cycle to be closed and thus greenhouse gas emissions to be reduced.

Besides organic wastes, ashes from biomass incineration are also worth considering for agricultural recycling. Wood energy production is classified as a form of green energy production because it is both carbon neutral and renewable (Kumar 2009). In many African regions, fuel wood constitutes 61–86% of primary energy consumption and generates large amounts of wood ash, which are just discarded to the natural environment without any control, causing serious environmental problems (Samir Amous 1999). On the other hand, in Europe and North America, the increased use of wood to produce bioenergy generates huge quantities of wood ash that are currently deposited at high cost. That ash, known to be rich in nutrients and lime, could be returned to depleted soils as a supplement to organic fertilizers by suitable management practices (Bougnom et al. 2009). Addition of wood ash to compost is among the available possibilities, and the compost produced could be used for forest fertilization as well as for agricultural purposes such as replenishing depleted and/or acid soils (Bougnom et al. 2009, 2010; Kuba et al. 2008). In this chapter, we explore the possibility of using both compost and wood ash as an additive to remediate tropical acid soils. The problem of soil acidity is also explained to inform the reader about the origin of this phenomenon and its consequence for soil fertility as well as the mechanisms by which organic wastes could alleviate soil acidity.

7.2 Soil Acidity

Soil acidity can be considered as the capacity of soils to manifest properties of acids or proton donors (Vorob'eva and Avdon'kin 2006). It occurs when acidity-generating processes outweigh acidity-consuming processes (Ulrich 1994). A soil is defined as acid when its pH is lower than 7. Soil acidification has many causes that are natural and unnatural. Although soil acidification is a slow natural process, it can be accelerated by plants, animals, and human activities or slowed down or reversed by careful management practices (Bolan et al. 1994; Poss et al. 1995).

7.2.1 Some Causes of Soil Acidity

7.2.1.1 Rainfall and Leaching

In climates where rainfall exceeds evapotranspiration, soils with low buffer capacity tend to acidify. Excess water infiltrating the soil enhances leaching of basic ions such as calcium (Ca^{2+}) and magnesium (Mg^{2+}) from the exchange complex of soil (clay minerals, humus) and their substitution by protons (H^+) and aluminum ions (Al^{3+}) (Mayer 1998). This way, neutral clay may be converted into a hydrogen clay or acid clay, which gradually accumulates and intensifies under increasing amounts of rainfall.

7.2.1.2 Acidic Parent Material

The parent material is the geological horizon from which soil horizons form; it is a key factor that in many cases determines the kinds and contents of secondary minerals of soils (Arbestain et al. 1999). Soils that develop from granite are likely to be more acidic than soils developed from calcareous limestone. In tropical and subtropical areas, under rainfall and high temperatures throughout the year, the process of acidification occurs over a long or a short time with the weathering of the soil parent material that liberates significant amounts of silica, iron, and aluminum and subsequent association of minerals of low crystallinity and aluminum–humus complexes (Garcia-Rodeja et al. 1987). In the humid tropics, most silicate minerals in the parent material are weathered away by desilication, leaving little other than the oxides of iron and aluminum (Sumner and Noble 2003).

7.2.1.3 Cropping Systems

Decay of organic matter, inappropriate use of nitrogenous fertilizers, and the removal of alkaline plant material from the field in cropping systems further accelerate soil acidification (Vieira et al. 2008). Some studies have shown that legumes increase soil acidification in pastures and arable cropping systems (Williams 1980; Burle et al. 1997). Acid production during carbon and nitrogen cycles is considered to be the most relevant in agricultural and pasture ecosystems (Helyar and Porter 1989). In non-polluted areas, soil acidification is mainly caused by the release of protons during the cycling of carbon, nitrogen, and sulfur in the soil–plant–animal system (Ulrich and Summer 1991). In a balanced system, carbon, nitrogen, and sulfur cycling processes are coupled and there is no generation of acidity. Perturbations such as uncoupling of nutrient cycles, accumulation of soil organic matter, leaching of nutrients (mainly NO_3^-) and the mobile exchangeable basic cations (Ca^{2+}, Mg^{2+}, K^+, and Na^+), and application of nitrogen fertilizers (which cause oxidation of NH_4^+ to NO_3^-) generate soil acidity. NO_3^- is not strongly adsorbed by the soil and will leach, and if it is not totally taken up by the crop, increased soil acidification will occur as a consequence (Bolan and Hedley 2003).

7.2.2 Problems of Acidic Soils

Acidification involves four main processes, which are (1) leaching of basic cations (Ca^{2+}, Mg^{2+}, K^+, and Na^+) from the exchangeable complex of the soil (clay, minerals, humus) and their substitution by protons (H^+) and cation acids ($Si_4{}^+$, $Fe_3{}^+$, $Al_3{}^+$), (2) accumulation of potentially toxic heavy metals (Co, Cd, Zn) in deeper soil layers, (3) accumulation of aluminum sulfates after saturation of exchange sites with Al ions that allows transportation of acidity to deeper soil layers or groundwaters, and (4) transfer of potentially toxic cation acids (Al, heavy metals) toward groundwater and surface water owing to the increased solubilization of compounds formed by these cation acids under very acidic conditions (Mayer 1998). Acid soil infertility is a major limitation to crop production on highly weathered and leached soils throughout the world (Von Uexküll and Mutert 1995). It is a complex interaction of growth-limiting factors including toxic levels of aluminum, manganese, and iron, as well as deficiencies of some essential elements, such as phosphorus, nitrogen, potassium, calcium, and magnesium, and some micronutrients (Kochian et al. 2004). Among these constraints, aluminum toxicity and phosphorous deficiency are the most important owing to their ubiquitous existence and overwhelming impact on plant growth (Kochian et al. 2004).

7.2.2.1 Aluminum Toxicity

Aluminum is a light metal that makes up 7% of the earth's crust and is the third most abundant element, after oxygen and silicon, plant roots are therefore almost always exposed to aluminum in some form (Ma et al. 2001). Aluminum exists in soils in many harmless forms, including hydrous oxides, aluminosilicates, sulfates, and phosphates (Haynes and Mokolobate 2001). Aluminum inhibits plant growth by affecting plant roots and development (Delhaize et al. 1993; Ma et al. 2001), inhibiting both cell divisions in the apical root meristem and cell elongation (Blamey et al. 1983). Roots become stubby and brittle; root tips and lateral roots become thick and occasionally necrotic brown (Mossor-Pietraszewska 2001). These effects restrict the ability of the plant to take up nutrients and water, leading to nutrient and/or water stresses (Rout et al. 2001; Haynes and Mokolobate 2001). Plants in acid soil, owing to aluminum solubility at low pH, exhibit a variety of nutrient-deficiency symptoms, with a consequent decrease in yield. Aluminum toxicity is linked with phosphorus, calcium, magnesium, or iron deficiency syndrome (Rout et al. 2001). To overcome the lack of productivity in aluminum toxic soil, the first step is to treat acidity, which will promote better root growth and function and will allow nutrients and water to be taken up more effectively by the plant.

7.2.2.2 Phosphorus Deficiency

In most humid tropical and subtropical regions where acid soils prevail, warm and moist conditions result in weathered soil types. These tropical soils are depleted in available phosphorus, and usually the total phosphorous level is low, ranging from 0.01 to 0.1% (Chen and Ma 2001). Much of the phosphorus is bound to aluminum and iron complexes during pedogenesis (Walker and Syers 1976). Through adsorption and precipitation mechanisms aluminum forms insoluble and stable complexes with inorganic and organic phosphates, forming highly insoluble phosphorous compounds; therefore, their solubilization is a prerequisite for phosphorous uptake by plants. Soils suffering from aluminum toxicity are generally associated with phosphorous deficiency. The phosphorous-use efficiency in such soils is around 10–15% in the best situations (Verma et al. 2005). The low phosphorous status of these soils is of great concern because large amounts of phosphorous need to be applied to raise the concentrations of available soil phosphorous to an adequate level (Sanchez and Uehara 1980). Low phosphorous availability is considered to be one of the main limiting factors to plant growth in acid soils, in addition to human wealth in tropical areas (Barber 1995; Sanchez 2002). For temperate soils, ash amendments have been shown to alleviate phosphorous deficiencies (see Chap. 2, Schiemenz et al. 2011).

7.2.3 Some Solutions to Soil Acidity

Since the problem of acid soil low fertility status has been known for long, many agricultural practices have been recommended to overcome the problem.

7.2.3.1 Liming

Liming is defined as the application of ground calcium and/or magnesium carbonates, hydroxides, and oxides. Liming the soil is the most common and oldest method for reducing soil acidity. Liming is often performed through high-dose applications of products such as calcitic lime ($CaCl_2$) and dolomitic lime [$CaMg(CO_3)_2$]. The aim is to increase the soil pH and therefore to modify the physical, chemical, and biological parameters of the soil. Studies have shown that liming materials affect the activity and composition of microbial populations and can create better environmental conditions for the development of nonacidophilic microorganisms, resulting in increased microbial biomass and soil respiration (Neale et al. 1997; Tate 2000). Nevertheless, liming has some limits; the effectiveness of surface application of lime to soils under a particularly no-till system withregard to subsoil acidity is uncertain, agricultural liming materials are relatively insoluble, and lime effects may be restricted to the top few centimeters of the soil surface for many years (Shainberg et al. 1989; Costa and Rosolem 2007).

Large quantities are generally required to improve plant growth, and for many resource-poor farmers in the tropics carrying out semisubsistence agriculture, its use is effectively prevented by the unavailability or the high cost of lime, or both (Haynes and Mokolobate 2001).

7.2.3.2 Use of Organic Waste Materials

Organic waste materials used to tackle soil acidity include undecomposed plant materials, composts, manures, peats, and coal products. Organic amendments are suitable for resource-poor farmers, as these farmers are unable to buy large quantities of lime and fertilizer phosphorous needed for their lands because of economic reasons. Some authors have reported an increase in soil pH after addition of organic materials to soil, followed by a decrease of aluminum saturation and an improvement of plant growth, depending on the type of residue, its rate of application, and the buffering capacity of the soil (Hue 1992; Noble et al. 1996). The rise of soil pH is due to the flow of protons from the soil (lower pH) to the organic matter sites (higher pH), decomposition of less stable materials in the soil resulting in mineralization and nitrification of organic nitrogen, and microbial decarboxylation (Haynes and Mokolobate 2001; Wong and Swift 2003). A long-term increase of soil pH is dependent on the balance between proton production and consumption in the system (Helyar and Porter 1989). The role of humic substances in increasing phosphorous availability is unclear. Some authors have reported the role of humic substances contained in organic matter in competing for adsorption soil sites and subsequent decrease in phosphorous adsorption (Bolan et al. 1994; Perrott 1978), whereas other authors have stated the unimportance of soil organic matter in increasing phosphorous availability (Borggaard et al. 1990). Humic substances concomitantly with organic acids, organic residues, and release of inorganic phosphorous have been found to be the main factors involved in increase of available phosphorous. A decrease in aluminum phytotoxicity is directly linked to phosphorous availability.

A conceptual model of the major processes that lead to the detoxification of soil aluminum and an increase of phosphorous availability when wood ash compost is applied to acid soils is summarized in Fig. 7.1.

7.3 Compost and Wood Ash

7.3.1 Compost

Compost is a product derived from composting that is high in nutrients and rich in humic acids. Nutrients are usually bound organically and thus released at a moderate speed (Gobat et al. 2003). Mature compost contains a diverse community of microorganisms; addition of compost to soil modifies considerably its biological,

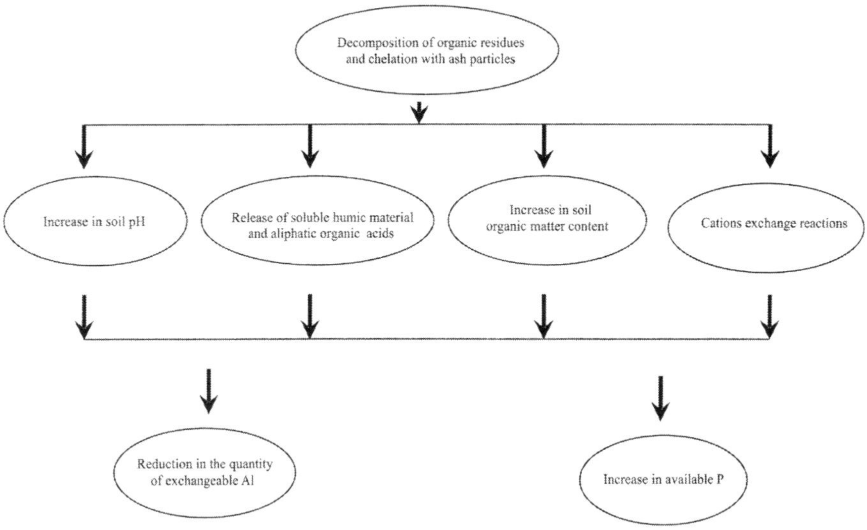

Fig. 7.1 A conceptual model of the major processes that lead to the detoxification of soil aluminum and an increase of phosphorous availability when wood ash compost is applied to acid soils

physical, and chemical properties in the short term as well as in the long term (Ryckeboer et al. 2003; Fuchs 2009). The use of compost in agriculture aids in replenishing and maintaining long-term soil fertility by providing optimal conditions for soil biological activity and a slow flow of nutrients adapted to the needs of the crop (Gobat et al. 2003).

7.3.1.1 Effect on Soil Physical Parameters

Depending on the amount, the type, the interval of application, and especially the characteristics of the soil, compost will improve soil structure and aggregate stability, hydraulic conductivity, infiltration, resilience against erosion, water holding capacity, air balance, and soil temperature (Gerzabek et al. 1995; Hartmann 2003). Stabilization of soil aggregates by organic matter occurs in three main ways: (1) Application of organic matter maintains the microbial activity and thereby the production of metabolic products with cementing properties by microbial degradation, mature compost performing better than immature compost; (2) application of organic matter supports the activity of the soil microfauna and mesofauna, e.g., earthworms. Excrement aggregates have positive effects on the soil structure and influence the formation of wide macropores and generally enhance microbial activity; (3) application of high molecular weight humic acids improves the long-term stability of microaggregates (Hartmann 2003; Fuchs et al. 2004). Annual applications of small amounts of compost are more effective in stabilizing aggregates and pores size distribution than any singular application of high amounts (Lamp 1996).

7.3.1.2 Effect on Soil Chemical Parameters

Composts (e.g., from green and kitchen waste) and soil differ in their composition and properties in almost every parameter. The amount of organic carbon, the amount of organic nitrogen, pH, electrical conductivity (EC), cation exchange capacity (CEC), salinity, chloride concentration, and sulfate concentration are higher in compost. Therefore, depending on the quality and quantity of compost, climate, and soil characteristics, their application will result in a modification of the soil organic matter composition (Fuchs et al. 2004). Compost application increases soil organic carbon and total nitrogen content at the upper horizon in both sandy and clay soils, and that increase can be observed even more than 10 years after application (Melero et al. 2007; Delschen 1999). Generally, an increase in soil pH after compost use is observed because most composts are basic and have a considerable buffering capacity. The mineralization of carbon and the subsequent production of OH^- ions by ligand exchange as well as the introduction of basic cations such as K^+, Ca^{2+}, and Mg^{2+} leads to an increase of soil pH (Mkhabela and Warman 2005). Compost application can ameliorate soil acidity by increasing soil pH and allows large quantities of lime required for these soils to be saved (Haynes and Mokolobate 2001). Application of municipal, mixed green and animal waste compost has been reported to proportionally increase the EC and salt effects of soils (Stamatiadis et al. 1999; Walter et al. 2006). Agricultural soil EC levels ranged from 0 to 4 dS m^{-1}, whereas soil EC levels of municipal solid waste composts ranged from 3.69 to 7.49 dS m^{-1} (Brady and Weil 1996); therefore, the increase of EC could be of concern. In some cases, soil EC levels were excessive and inhibited plant growth; nevertheless, Zhang et al. (2006) reported that the increased soil EC values decline over time because of nutrient removal by crops and leaching, but the long-term soil biological activity could be negatively affected (Iglesias-Jimenez and Alvarez 1993). Incorporation of compost into soil, especially at high doses, increases the CEC (Bengtson and Cornette 1973); the rise is generally linked to an increase in the level of organic material, the pH, and in base saturation (Ca, K, Na) (Fuchs et al. 2004). Nitrate leaching is one of the concerns regarding compost utilization. NO_3^- is highly mobile in soils and is susceptible to leaching through the soil profile and into the groundwater by the infiltrating water. When compared with organic and mineral fertilizers, the nitrate leaching potential of composts is very low (Insam and Merschak 1997). Composts for agricultural use should come from source-separated organic waste and green waste only; then heavy metals are not of concern, as they would be if municipal solid waste compost were used (Epstein et al. 1992; Sharma et al. 1997).

7.3.1.3 Effect on Soil Biological Parameters

Effect on Soil Microorganisms

Heterotrophic organisms in the soil are ultimately responsible for ensuring the availability of nutrients for primary production (Wardle 2002). Microorganisms play a very important role in many biogeochemical cycles in agroecosystems including organic matter decomposition, nutrient mineralization, and trace gas emission and consumption (Carney et al. 2004). The principal "players" in the decomposition process are microorganisms, i.e., bacteria, archaea, and fungi.

Bacteria are able to perform an extremely wide range of chemical transformations, but are, however, only active over a very narrow range of environmental conditions (Lavelle and Spain 2001). As with all microorganisms, bacteria have a system of external digestion mediated through the production of extracellular enzymes, and some of the metabolites released by extracellular digestion may be used by other organisms, thus creating a trophic stimulus for opportunistic or cooperating microorganisms (Hattori 1973; Lavelle and Spain 2001). Until recently, archaea were considered to occur in extreme environments only, but their presence was also reported in numerous other habitats, including forest and agricultural soils, where their potential for ammonia oxidation was demonstrated (Bintrim et al. 1997; Pace 1997; Prosser and Nicol 2008). Bacteria and archaea, on the one hand, and fungi, on the other, differ biochemically and morphologically. Fungi are larger than bacteria and have hyphae that can grow into and explore distant microhabitats, and translocate carbon and nitrogen and other nutrients within the hyphal network. Thus, fungi are regarded as being more capable than bacteria and actinobacteria in degrading polysaccharides (Atlas and Bartha 1998; Lavelle and Spain 2001). The broad functions of fungal mycelium in soil and litter are decomposition and nutrient cycling. In contrast to bacteria, fungi can remain active in soils at very low water potential (−7,200 kPa) and are better suited than bacteria to exist in interpore spaces (Shipton 1986). These microorganisms influence or control ecosystem processes and form mycorrhizal interactions with plants (Coleman 2001; Wardle 2002).

Soil microbial community diversity has been suggested as a way of assessing the "health" or "quality" of soils (Chapman et al. 2007). High biodiversity may be vitally important in structurally diverse ecosystems such as soil because it may promote productivity and stability of this environment (Grime 1997; van Bruggen and Semenov 2000). The biodiversity of fungal or bacterial populations in the rhizosphere is closely related to growth of crops; hence, crop yield may be used as an indicator of soil health associated with greater stability in productivity (Lynch et al. 2004).

The effect of different composts on the microbial biomass and diversity depends in part on the amount used and very strongly on the compost quality (Ros et al. 2006). Populations of rhizosphere microorganisms were reported to increase in relation to increasing inputs of composted organic matter to soil, and compost application has been found to enhance biomass nitrogen, carbon, and sulfur content and microbial activity over several years (Perucci 1990; Ros et al. 2006). Single and repetitive applications of different amounts of organic wastes significantly increase the amount of soil microbial biomass and enhance nitrogen mineralization potential, but excessive rates of application (100 t ha^{-1}) reduce the functional diversity of the microbial community (Banerjee et al. 1997). Several studies have reported modification of both bacterial and fungal community structure following application of compost (Crecchio et al. 2004; Ros et al. 2006; Innerebner et al. 2006). Increases in dehydrogenase, β-glucosidase, urease, nitrate reductase, and phosphatase activities were observed 3 months after application of municipal solid waste compost (Crecchio et al. 2004). Some composts rich in heavy metals (Zn, Cu, and Pb) have been reported to decrease enzyme (phosphatase and urease) activities, whereas other enzymes (dehydrogenase, catalase, protease) were not affected (Garcia-Gil et al. 2000).

Effects on Soil Fauna

The relative contributions of the soil fauna to microbial turnover and nutrient mineralization are directly related to the demographics of the soil biota (Coleman et al. 1983). Many invertebrates (e.g., earthworms, termites, and ants) play an important role for soil fertility; they produce macropores (e.g., galleries, chambers) and organomineral structures that influence hydraulic properties, macroaggregation, and organic matter dynamics in soil (Lavelle 1997, 2002). The importance of earthworms in enhancing nutrient availability and raising the rate of nitrogen turnover through the breakdown and incorporation of organic matter into the soil was demonstrated (Basker et al. 1992). Leroy et al. (2007) and Moreira et al. (2008) reported a significant increase of earthworm density and biomass after compost application. Earthworms can affect other soil-inhabiting invertebrates by altering their resource base, affecting soil structure, by direct ingestion, and by dispersing them (Blair et al. 1995); a direct effect of earthworm activities following compost application is an increase in the number of trophic groups of soil-inhabiting arthropods (phytophages, predators, omnivores, and saprovores) (Gunadi et al. 2002). Nematodes play a role in decomposition and nutrient cycling; free-living nematodes that feed on bacteria and fungi (as opposed to plants) contribute as much as 27% of the readily available nitrogen in the soil and promote rhizosphere colonization of beneficial rhizobacteria (Ekschmitt et al. 1999; Knox et al. 2003). Compost supply stimulates nonparasitic nematodes and encourages predator nematodes and parasitic fungi which specifically destroy the eggs of certain parasitic nematodes (Fuchs et al. 2004). By boosting the key functional species in soil defined as “ecosystem engineers” (Jones et al. 1994), compost allows optimal conditions for plant health and growth.

7.3.2 Wood Ash

Wood ash is a by-product of the wood industry resulting from burning of wood residues for energy production (Nkana et al. 2002). Most of the inorganic nutrients and trace elements in wood are retained in the ash during combustion; the quality of the end product depends on the quality of the wood, the tree species, and the burning process (Perkiömäki et al. 2004). The ash and the metal contents are generally higher in bark than in stemwood (Hakkila 1986; Werkelin et al. 2005). Fly ash contains higher levels of dioxins and heavy metals than the bottom ash (Pitman 2006). For agricultural and horticultural purposes, only bottom ash should be used according to Stockinger et al. (2006). Wood ash is a significant source of the nutrients phosphorous, potassium, magnesium, and calcium, and its properties resemble those of lime (Naylor and Schmidt 1986; Ohno and Erich 1990). Ash fertilization can compensate for the nutrient losses caused by harvesting operations, nutrient leaching, and soil acidification (Saarsalmi et al. 2006).

7.3.2.1 Effect on Soil Physical Parameters

More than 80% of wood ash is composed of particles smaller than 1.0 mm, with all particle sizes ranging from coarse sand to clay (Etiegni et al. 1991; Etiegni and Campbell 1991). Wood ash particles may swell in contact with water and can clog soil pores and therefore affect the texture, decrease aeration, and increase water holding capacity and hydraulic conductivity of the soil (Etiegni and Campbell 1991). Ash-amended soils tend to have lower bulk density, higher water holding capacity, lower hydraulic conductivity, and increase the protection of soil against erosion (Eisenberg et al. 1986; Kalra et al. 1998). Application of ash for agricultural purposes should be done with precaution because of the detrimental effects observed at high rates; amounts of ash exceeding 20% (v/v) in calcareous soils and 10% in acidic soils have been reported to deteriorate the hydraulic conductivity (Gupta et al. 2002). However, many of the available studies refer to fly ash of coal, which has distinctive properties different from those the wood ash we are dealing with.

7.3.2.2 Effect on Soil Chemical Parameters

Calcite ($CaCO_3$) and fairchildite [$K_2Ca(CO_3)_2$] at a combustion temperature of 600°C and lime (CaO) and magnesium oxide (MgO) at a combustion temperature of 1,300°C are the major compounds of wood ash (Etiegni and Campbell 1991; Misra et al. 1993). Wood ash alkalinity or neutralizing capacity expressed as calcium carbonate equivalent ranges from less than 20% to over 90% (Vance 1996). It typically ranges from 50 to 100% of that of limestone on a dry weight basis (Naylor and Schmidt 1986; Lerner and Utzinger 1986). The soluble alkalinity is mainly due to hydroxides (92%) and a small amount of carbonates (8%) of calcium, magnesium, and potassium (Etiegni and Campbell 1991). The pH (H_2O) of wood ash ranges between 8.9 and 13.5 (Demeyer et al. 2001) and when wood ash is applied, it acts faster in raising soil pH than lime: the pH increase is higher for soils with low pH and low organic matter content (Ohno 1992). The oxides, hydroxides, and carbonates (quickly soluble) are responsible for the faster soil pH rise, which does not persist for a long time in the soil, whereas calcite (less soluble) lasts longer and is responsible for the moderate alkalinity of the soil (Ulery et al. 1993;

Muse and Mitchell 1995). Many studies (Nkana et al. 1998) have shown that wood ash appears to be a neutralizer of soil acidity and lowers aluminium and manganese toxicity in acid soils. Application of wood ash increases the CEC (Eriksson 1998; Nkana et al. 1998). Wood ash contains very little carbon and nitrogen because of their oxidization and transformation to gaseous constituents during combustion (Demeyer et al. 2001). However, it can lead to an increase of the organic carbon solubility and nitrification rate and potentially an increase of the nitrogen leaching rate (Pietikäinen and Fritze 1995; Kahl et al. 1996). However, nitrogen availability often increases indirectly as a result of ash application owing to a rise in soil pH and consequent nitrogen mineralization (Pitman 2006). The wood ash fertilization effect has been shown to act in a way similar to that of NPK commercial fertilizers on the basis of the nitrogen, phosphorous (P_2O_5), and potassium (K_2O) concentrations in the ratio 0:1:3 for a commercial wood boiler and 0:3:14 for a domestic wood stove (lower temperature burn) (Naylor and Schmidt 1986).

Wood ash constitutes a significant source of nutrients such as calcium, potassium, magnesium, silicon, and phosphorous as the major macroelements (Etiegni and Campbell 1991; Vance 1996). Because of the complicated nature of wood ash (several cations forming oxides, hydroxides, carbonates, and bicarbonates), the dissolution rates of the compounds formed are different. Potassium dissolves very quickly (more than 50% of the total), calcium and magnesium dissolve more rapidly with decreasing ash-to-water ratio, and phosphorous remains relatively insoluble and may be immobilized as iron and aluminum phosphates in acidic soils (Ohno 1992; Khanna et al. 1994). Wood ash contains many micronutrients (Fe, Mn, Zn, Cu, B, Mo, Cr, Pb, Co, Cd), with iron as the most abundant, followed by manganese, zinc and copper, boron, and molybdenum. Depending on the soil pH, application of wood ash leads to lower iron, manganese, zinc, and copper availability at pH ≥ 6.5; if the pH decreases, the micronutrients become more mobile and their concentrations in soil increase (Clapham and Zibilske 1992; Zhan et al. 1996). Heavy metals hazardous to human health, such as arsenic, cadmium, chromium, lead, mercury, and selenium, and those that may be phytotoxic, but are not normally hazardous to humans, such as boron, copper, nickel, and zinc (ICRCL 1987), are generally found in wood ash and could be of concern, but the heavy metal concentration in agricultural fertilizers frequently exceeds that from bottom wood boiler ash. Variability of zinc, nickel and copper contents is of concern only if large quantities of wood ash are used; chromium, cadmium, arsenic, mercury, and lead contents are harmless as long as fly ash is not used (Pitman 2006).

7.3.2.3 Effect on Soil Biological Parameters

Effect on Soil Microorganisms

Wood ash may have a stimulatory or an inhibitory effect (when applied at high dose, e.g., more than 5 Mg ha^{-1}), or an initial stimulation and later an inhibition effect on microbial activity and biomass (Baath et al. 1995; Zimmermann and Frey 2002).

The changes in microbial biomass and microbial activity are related to the increase in soil pH, which induces the development of bacteria to the detriment of fungi (Perucci et al. 2006). Modifications of both bacterial and fungal community structure have also been reported following wood ash application, with the fungal community being more sensitive to the ash dose (Baath et al. 1995). Alkaline phosphatase, arylsulfatase, and catalase activities increased under wood ash application at 5 t ha^{-1}, but were affected negatively at 20 t ha^{-1} (Perucci et al. 2006). Generally, the different changes in microbial biomass and composition are related to the dose of ash applied (Takeuchi et al. 2009).

Effects on Soil Fauna

Wood ash application has generally shown little effect on soil fauna (Haimi et al. 2000; Liiri et al. 2002a). The effect of wood ash application on enchytraeids or earthworms depending on the rate of application has shown to be positive (Huhta et al. 1986), neutral (Lundkvist 1998), or negative (Haimi et al. 2000). When wood ash was applied at a rate 3 Mg ha^{-1}, there was no change in the number of microarthropods (Liiri et al. 2002a), whereas their total number decreased at an application rate of 5 Mg ha^{-1} (Haimi et al. 2000) in a pine forest. A positive correlation was found between wood ash treatment and an increase in the number of nematodes, especially of bacterial feeders (Huhta et al. 1986; Liiri et al. 2002a). However, a shift from fungi to bacteria might hamper number of enchytraeids (see Chap. 4, Nieminen 2011). The impact of wood ash on mites and collembola was positive, neutral, or negative (Huhta et al. 1986; Haimi et al. 2000; Liiri et al. 2002b). Once more, the dose was the main factor responsible for these results. Wood ash has a negative effect on the soil fauna if large quantities are used (Siddiqui and Singh 2005).

7.4 Conclusions

Regarding the attributes of these wastes, co-composting organic wastes with wood ash as an additive to obtain a compost with higher liming potential and rich in nutrients is of great interest for both solving the problem of waste disposal and providing an economically attractive alternative to chemical fertilizers for acid tropical soils. Compost with wood ash additive could be an excellent amendment to compensate for loss of nutrients or to treat soil acidity, or both. Preliminary studies have already been carried out using wood ash compost in greenhouse experiments by Bougnom et al. (2009, 2010). The results of these experiments revealed that the use of compost with wood ash as an additive presents no risk to a sustainable agriculture in the tropics in terms of plant toxicity, physical and chemical parameters, and impact on the soil microbiota if a small amount of ash is added. Wood ash compost represents new chances in tackling infertility of acid tropical soils as it could be an efficient and yet affordable alternative to lime. Organic waste composts are available in very limited amounts in the tropics and despite fuel wood constituting the primary energy consumption, little ash for large-scale application is available. Balanced ash compost could allow the amounts of organic waste compost needed to reach suitable pHs to be reduced, restore soil organic matter pools, and remediate nutrient deficiencies in agricultural soils. Cautiously prepared and organized, the use of high-quality wood ash, otherwise discarded in Europe, if substituted to some mineral feedstock, could be an option for ameliorating acid tropical soils, hitting two birds with one stone: less has to be deposited in waste dumps and impoverished tropical soils might be improved.

References

Arbestain MC, Barreal ME, Macías F (1999) Parent material influence on sulfate sorption in forest soils from Northwestern Spain. Soil Sci Soc Am J 63:1906–1914

Atlas RM, Bartha R (1998) Microbial ecology: fundamentals and application, 4th edn. Benjamin Cummings, San Francisco, p 694

Baath E, Frostegard A, Pennanen T, Fritze H (1995) Microbial community structure and pH response in relation to soil-organic-matter quality in wood-ash fertilized, clear-cut or burned coniferous forest soils. Soil Biol Biochem 27:229–240

Banerjee MR, Burton DL, Depoe S (1997) Impact of sewage sludge application on soil biological characteristics. Agric Ecosyst Environ 66:241–249

Barber SA (1995) Soil nutrient bioavailability: a mechanistic approach. Wiley, New York, pp 202–230

Basker A, Macgregor AN, Kirkman JH (1992) Influence of soil ingestion by earthworms on the availability of potassium in soil: an incubation experiment. Biol Fertil Soils 14:300–303

Bengtson GW, Cornette JJ (1973) Disposal of composted municipal waste in a plantation of young slash pine: effects on soil and trees. J Environ Qual 2:441–444

Bintrim SB, Donohue TJ, Handelsman J, Roberts GP, Goodman RM (1997) Molecular phylogeny of archaea from soil. Proc Natl Acad Sci USA 94:277–282

Blair JM, Parmelee RW, Lavelle P (1995) Influences of earthworms on biogeochemistry. In: Hendrix PF (ed) Earthworm ecology and biogeography in North America. Lewis, Boca Raton, pp 127–158

Blamey FPC, Edwards DG, Asher CJ (1983) Effects of aluminum, OH-Al and P-Al molar ratios, and ionic strength on soybean root elongation in solution culture. Soil Sci 136:197–207

Bolan NS, Hedley MJ (2003) Role of carbon, nitrogen, and sulfur cycles in soil acidification. In: Rengel Z (ed) Handbook of soil acidity. Marcel Dekker, New York, pp 29–56

Bolan NS, Naidu R, Mahimairaja S, Baskaran S (1994) Influence of low-molecular weight organic acids on the solubilization of phosphates. Biol Fertil Soils 18:311–319

Borggaard OK, Jorgensen SS, Moberg JP, Rabenlange B (1990) Influence of organic-matter on phosphate adsorption by aluminium and iron oxides in sandy soils. J Soil Sci 41:443–449

Bougnom BP, Maier J, Etoa FX, Insam H (2009) Composts with wood ash addition: a risk or a chance for ameliorating acid tropical soils? Geoderma 153:402–407

Bougnom BP, Knapp BA, Etoa FX, Elhottová D, Němcová A, Insam H (2010) Designer composts for ameliorating acid tropical soils: effects on the soil microbiota. Appl Soil Ecol 45:319–324

Brady N, Weil R (1996) The nature and properties of soils, 12th edn. Prentice, Upper Saddle River, New Jersey, p 385, 495

Burle ML, Mielniczuk J, Focchi S (1997) Effect of cropping systems on soil chemical characteristics, with emphasis on soil acidification. Plant Soil 190:309–316

Carney KM, Matson PA, Bohannan BJM (2004) Diversity and composition of tropical soil nitrifiers across a plant diversity gradient and among land-use types. Ecol Lett 7:684–694
Chapman SJ, Campbell CD, Artz RRE (2007) Assessing CLPPs using MicroResp$^{(TM)}$. A comparison with biolog and multi-SIR. J Soil Sediment 7:406–410
Chen M, Ma LQ (2001) Taxonomic and geographic distribution of total phosphorus in Florida surface soils. Soil Sci Soc Am J 65:1539–1547
Clapham WM, Zibilske LM (1992) Wood ash as a liming amendment. Commun Soil Sci Plant Anal 23:1209–1227
Coleman DC (2001) Soil biota, soil systems and processes. In: Levin S (ed) Encyclopedia of biodiversity, vol 5. Academic, San Diego, pp 305–314
Coleman DC, Reid CPP, Cole CV (1983) Biological strategies of nutrient cycling in soil systems. Adv Ecol Res 13:1–55
Costa A, Rosolem CA (2007) Liming in the transition to no-till under a wheat soybean rotation. Soil Tillage Res 97:207–217
Crecchio C, Curci M, Pizzigallo MDR, Ricciuti P, Ruggiero P (2004) Effects of municipal solid waste compost amendments on soil enzyme activities and bacterial genetic diversity. Soil Biol Biochem 36:1595–1605
Delhaize E, Ryan PR, Randall PJ (1993) Aluminum tolerance in wheat (*Triticum aestivum* L.) II: Aluminum-stimulated excretion of malic acid from root apices. Plant Physiol 103:695–702
Delschen T (1999) Impacts of long-term application of organic fertilizers on soil quality parameters in reclaimed loess soils of the Rhineland lignite mining area. Plant Soil 213:43–54
Demeyer A, Nkana JCV, Verloo MG (2001) Characteristics of wood ash and influence on soil properties and nutrient uptake: an overview. Bioresour Technol 77:287–295
Eisenberg SH, Tittlebaum ME, Eaton HC, Soroczak MM (1986) Chemical characteristics of selected fly ash leachate. J Environ Sci Health A 21:383–402
Ekschmitt K, Bakonyi G, Bongers M, Bongers T, Bostrom S, Dogan H, Harrison A, Kallimanis A, Nagy P, O'Donnell AG, Sohlenius B, Stamou GP, Wolters V (1999) Effects of the nematofauna on microbial energy and matter transformation rates in European grassland soils. Plant Soil 212:45–61
Epstein E, Chaney RL, Henry C, Logan TJ (1992) Trace elements in municipal solid waste compost. Biomass Bioenergy 3:227–238
Eriksson HM (1998) Short-term effects of granulated wood ash on forest soil chemistry in SW and NE Sweden. Scand J For Res 43–55
Etiegni L, Campbell AG (1991) Physical and chemical characteristics of wood ash. Bioresour Technol 37:173–178
Etiegni L, Campbell AG, Mahler RL (1991) Evaluation of wood ash disposal on agricultural land: I. Potential as a soil additive and liming agent. Commun Soil Sci Plant Anal 243–256
Fuchs JG (2009) Interactions between beneficial and harmful microorganisms: from the composting process to compost application. In: Insam H, Franke-Whittle IH, Goberna M (eds) From waste to resource: microbes do the job. Microbes at work. Springer, Berlin, pp 213–230
Fuchs JG, Bieri M, Chardonnens M (2004) Effects of compost and digestate on the environment, soil fertility and plant health. In: Fuchs JG, Bieri M (eds) Review of the current literature. FiBL, Frick, pp 1–16
Garcia-Gil JC, Plaza C, Soler-Rovira P, Polo A (2000) Long-term effects of municipal solid waste compost application on soil enzyme activities and microbial biomass. Soil Biol Biochem 32:1907–1913
Garcia-Rodeja E, Silva BM, Macıas F (1987) Andosols developed from non-volcanic materials in Galicia, NW Spain. J Soil Sci 38:573–591
Gerzabek MH, Kirchmann H, Pichlmayer F (1995) Response of soil aggregate stability to manure amendments in the Ultuna long- term soil organic matter experiment. Z Pflanzenernahr Dung Bodenkd 158:257–260
Gobat JM, Aragno M, Mattey W (2003) Le sol vivant, 2e édition revue et complétée. Presses Polytechniques et Universitaires Romandes, Lausanne, p 319

Grime JP (1997) Ecology – biodiversity and ecosystem function: the debate deepens. Science 277:1260–1261

Gunadi B, Edwards CA, Arancon NQ (2002) Changes in trophic structure of soil arthropods after the application of vermicomposts. Eur J Soil Biol 38:161–165

Gupta DK, Rai UN, Tripathi RD, Inouhe M (2002) Impacts of fly-ash on soil and plant responses. J Plant Res 115:401–409

Haimi J, Fritze H, Moilanen P (2000) Responses of soil decomposer animals to wood ash fertilisation and burning in a coniferous forest stand. For Ecol Manage 129:53–61

Hakkila P (1986) Recycling of wood and bark ash. State-of-the-art review. IEA Forest Energy Agreement, Helsinki

Hartmann R (2003) Studien zur standortgerechten Kompotanwendung, vol 39, Bremer Beiträge zur Geographie und Raumplanung. Universität Bremen, Bremen

Hattori T (1973) Microbial life in the soil. Marcel Dekker, New York

Haynes RJ, Mokolobate MS (2001) Amelioration of Al toxicity and P deficiency in acid soils by additions of organic residues: a critical review of the phenomenon and the mechanisms involved. Nutr Cycl Agroecosyst 59:47–63

Helyar KR, Porter WM (1989) Soil acidification its measurement and the processes involved. In: Robson AD (ed) Soil acidity and plant growth. Academic, Sydney, pp 61–101

Hue NV (1992) Correcting soil acidity of a highly weathered ultisol with chicken manure and sewage sludge. Commun Soil Sci Plant Anal 23:241–264

Huhta V, Hyvönen R, Koskenniemi A, Vilkamaa P, Kaasalainen P, Sulander M (1986) Response of soil fauna to fertilization and manipulation of pH in coniferous forests. Acta For Fenn 195:1–30

Iglesias-Jimenez E, Alvarez CE (1993) Apparent availability of nitrogen in composted municipal refuse. Biol Fertil Soils 16:313–318

Innerebner G, Knapp B, Vasara T, Romantschuk M, Insam H (2006) Traceability of ammonia-oxidizing bacteria in compost-treated soils. Soil Biol Biochem 38:1092–1100

Insam H, Merschak P (1997) Nitrogen leaching from forest soil cores after amending organic recycling products and fertilizers. Waste Manage Res 15:277–291

Interdepartmental Committee on the Redevelopment of Contaminated Land (ICRCL) (1987) Guidance on the assessment and redevelopment of contaminated land. Guidance note 59/83. Department of the Environment, London

Jones CG, Lawton JH, Shachak M (1994) Organisms as ecosystem engineers. Oikos 69:373–386

Kahl JS, Fernandez IJ, Rustad LE, Peckenham J (1996) Threshold application rates of wood ash to an acidic forest soil. J Environ Qual 25:220–227

Kalra N, Jain MC, Joshi HC, Choudhary R, Harit RC, Vatsa BK, Sharma SK, Kumar V (1998) Fly ash as a soil conditioner and fertilizer. Bioresour Technol 64:163–167

Khanna PK, Raison RJ, Falkiner RA (1994) Chemical properties of ash derived from Eucalyptus litter and its effects on forest soils. For Ecol Manage 66:107–125

Knox OGG, Killham K, Mullins CE, Wilson MJ (2003) Nematode-enhanced microbial colonization of the wheat rhizosphere. FEMS Microbiol Lett 225:227–233

Kochian LV, Hoekenga OA, Pineros MA (2004) How do crop plants tolerate acid soils? – mechanisms of aluminum tolerance and phosphorous efficiency. Annu Rev Plant Biol 55 (459):493

Kuba T, Tschöll A, Partl C, Meyer K, Insam H (2008) Wood ash admixture to organic wastes improves compost and its performance. Agric Ecosyst Environ 27:43–49

Kumar A (2009) A conceptual comparison of bioenergy options for using mountain pine beetle infested wood in Western Canada. Bioresour Technol 100:387–399

Lamp D (1996) Wirkung von Müllklärschlamm- und Bioabfallkompost auf physikalische Bodeneigenschaften in mehrjährigen Feldversuchen auf schluffigen Lehmböden. Lehrstuhl für Pflanzenernährung, TU München-Weihenstephan, Freising-Weihenstephan, p 58

Lavelle P (1997) Faunal activities and soil processes: adaptive strategies that determine ecosystem function. Adv Ecol Res 27:93–132

Lavelle P (2002) Functional domains in soils. Ecol Res 17:441–450
Lavelle P, Spain AV (2001) Soil ecology. Kluwer, Dordrecht, p 654
Lerner BR, Utzinger JD (1986) Wood ash as a soil liming material. Hortscience 21(1):76–78
Leroy BLM, Van den Bossche A, NeveS D, Reheul D, Moens M (2007) The quality of exogenous organic matter: short-term influence on earthworm abundance. Eur J Soil Biol 43:S196–S200
Liiri M, Haimi J, Setala H (2002a) Community composition of soil microarthropods of acid forest soils as affected by wood ash application. Pedobiologia 46:108–124
Liiri M, Setala H, Haimi J, Pennanen T, Fritze H (2002b) Soil processes are not influenced by the functional complexity of soil decomposer food webs under disturbance. Soil Biol Biochem 34:1009–1020
Lundkvist H (1998) Wood ash effects on enchytraeid and earthworm abundance and enchytraeid cadmium content. Scand J For Res 2:86–95
Lynch J, Benedetti A, Insam H, Smalla C, Torsvik V, Nuti M, Nannipieri P (2004) Microbial diversity in soil: ecological theories, the contribution of molecular techniques and the impact of transgenic plants and transgenic microorganisms – a review. Biol Fertil Soils 40:363–385
Ma JF, Ryan PR, Delhaize E (2001) Aluminium tolerance in plants and the complexing role of organic acids. Trends Plant Sci 6:273–278
Mayer R (1998) Soil acidification and cycling of metal elements: cause-effect relationships with regard to forestry practices and climatic changes. Agric Ecosyst Environ 67:145–152
Melero S, Madejon E, Ruiz JC, Herencia JF (2007) Chemical and biochemical properties of a clay soil under dryland agriculture system as affected by organic fertilization. Eur J Agron 26:327–334
Misra MK, Ragland KW, Baker AJ (1993) Wood ash composition as a function of furnace temperature. Biomass Bioenergy 4:103–116
Mkhabela MS, Warman PR (2005) The influence of municipal solid waste compost on yield, soil phosphorus availability and uptake by two vegetable crops grown in a Pugwash sandy loam soil in Nova Scotia. Agric Ecosyst Environ 106:57–67
Moreira R, Sousa JP, Canhoto C (2008) Biological testing of a digested sewage sludge and derived composts. Bioresour Technol 99:8382–8389
Mossor-Pietraszewska T (2001) Effect of aluminium on plant growth and metabolism. Acta Biochim Pol 3:673–686
Muse JK, Mitchell CC (1995) Paper mill boiler ash and lime by-products as soil liming materials. Agron J 87:432–438
Naylor LM, Schmidt EJ (1986) Agricultural use of wood ash as a fertiliser and liming material. TAPPI J 69:114–119
Neale SP, Shah Z, Adams WA (1997) Changes in microbial biomass and nitrogen turnover in acidic organic soils following liming. Soil Biol Biochem 29:1463–1474
Nieminen JK (2011) Wood ash effects on soil fauna and interactions with carbohydrate supply: a mini-review. In: Insam H, Knapp BA (eds) Recycling of biomass ashes. Springer, Heidelberg, pp 45–56
Nkana JCV, Demeyer A, Verloo MG (1998) Chemical effects food ash on plant growth in tropical acid soils. Bioresour Technol 63:251–260
Nkana JCV, Demeyer A, Verloo MG (2002) Effect of wood ash application on soil solution chemistry of tropical acid soils: incubation study. Bioresour Technol 85:323–325
Noble AD, Zenneck I, Randall PJ (1996) Leaf litter ash alkalinity and neutralisation of soil acidity. Plant Soil 179:293–302
Ohno T (1992) Neutralization of soil acidity and release of phosphorus and potassium by wood ash. J Environ Qual 21:433–438
Ohno T, Erich MS (1990) Effect of wood ash application on soil pH and soil test nutrient levels. Agric Ecosyst Environ 32:223–239
Pace NR (1997) A molecular view of microbial diversity and the biosphere. Science 276:734–740

Perkiömäki J, Levula T, Fritze H (2004) A reciprocal decomposition experiment of Scots pine needles 19 yr after wood ash fertilization. Soil Biol Biochem 36:731–734

Perrott KW (1978) The influence of organic matter extracted from humified clover on the properties of amorphous aluminosilicates. I. Surface charge. Aust J Soil Res 16:327–339

Perucci P (1990) Effect of the addition of municipal solid-waste compost on microbial biomass and enzyme activities in soil. Biol Fertil Soils 10:221–226

Perucci P, Monaci E, Casucci C, Vischetti C (2006) Effect of recycling wood ash on microbiological and biochemical properties of soils. Agron Sustain Dev 26:157–165

Pietikäinen J, Fritze H (1995) Clear-cutting and prescribed burning in coniferous forest: comparison of effects on soil fungal and total microbial biomass, respiration activity and nitrification. Soil Biol Biochem 27:101–109

Pitman RM (2006) Wood ash use in forestry – a review of the environmental impacts. Forestry 79:563–588

Poss R, Smith CJ, Dunin FX, Angus JF (1995) Rate of soil acidification under wheat in a semi arid environment. Plant Soil 177:85–100

Prosser JI, Nicol GW (2008) Relative contributions of archaea and bacteria to aerobic ammonia oxidation in the environment. Environ Microbiol 10:2931–2941

Rigby D, Caceres D (2001) Organic farming and the sustainability of agricultural systems. Agric Syst 68:21–40

Ros M, Klammer S, Knapp B, Aichberger K, Insam H (2006) Long-term effects of compost amendment of soil on functional and structural diversity and microbial activity. Soil Use Manage 22:209–218

Rout GR, Samantaray S, Das P (2001) Aluminium toxicity in plants: a review. Agronomie 21:3–21

Ryckeboer J, Mergaert J, Vaes K, Klammer S, De Clercq D, Coosemans J, Insam H, Swings J (2003) A survey of bacteria and fungi occurring during composting and self-heating processes. Ann Microbiol 53:349–410

Saarsalmi A, Kukkola M, Moilanen M, Arola M (2006) Long-term effects of ash and N fertilization on stand growth, tree nutrient status and soil chemistry in a Scots pine stand. For Ecol Manage 235:116–128

Samir Amous (1999) The role of wood energy in Africa. In: Rivero SI (ed) Wood energy today for tomorrow. Food and Agriculture Organization of the United Nations, Rome. http://www.fao.org/docrep/x2740E/x2740e00.htm#acro. Accessed 11 Apr 2010

Sanchez PA (2002) Soil fertility and hunger in Africa. Science 295:2019–2020

Sanchez PA, Uehara G (1980) Management considerations for acid soils with high phosphorus fixation capacity. In: Khasawneh FE, Sample EC, Kamprath EJ (eds) The role of phosphorus in agriculture. ASA, Madison, pp 471–509

Schiemenz K, Kern J, Paulsen HP, Bachmann S, Eichler-Löbermann B (2011) Phosphorus fertilizing effects of biomass ashes. In: Insam H, Knapp BA (eds) Recycling of biomass ashes. Springer, Heidelberg, pp 17–31

Shainberg I, Sumner ME, Miller WP, Farina MPW, Pavan MA, Fey MV (1989) In: Stewart BA (ed) Use of gypsum on soils: a review, vol 9, Advanced in soil science. Springer, New York, pp 1–111

Sharma VK, Canditelli M, Fortuna F, Cornacchia G (1997) Processing of urban and agro industrial residues by aerobic composting: review. Energy Convers Manage 38:453–478

Shipton PJ (1986) Infection by foot and root rot pathogens and subsequent damage. In: Wood RKS, Jellis GK (eds) Plant diseases, infection, damage and loss. Blackwell, Oxford, pp 139–155

Siddiqui ZA, Singh LP (2005) Effects of fly ash, *Pseudomonas striata* and *Rhizobium* on the reproduction of nematode *Meloidogyne incognita* and on the growth and transpiration of pea. J Environ Biol 26:117–122

Stamatiadis S, Werner M, Buchanan M (1999) Field assessment of soil quality as affected by compost and fertilizer application in a broccoli field (San Benito County, California). Appl Soil Ecol 12:217–225

Stockinger M, Kuba T, Plank R, Meyer K, Insam H (2006) Die Verwertung von Holzaschen ein lösbares Problem? Österr Forstztg 117:20–21
Sumner ME, Noble AD (2003) Soil acidification: the world story. In: Rengel Z (ed) Handbook of soil acidity. Marcel Dekker, New York, p 12
Takeuchi M, Kawahata H, Gupta LP, Itouga M, Sakakibara H, Ohta H, Komai T, Ono Y (2009) Chemistry of fly ash and cyclone ash leachate from waste materials and effects of ash leachates on bacterial growth, nitrogen-transformation activity, and metal accumulation. J Hazard Mater 165:967–973
Tate RL (2000) Soil microbiology, 2nd edn. Wiley, New York
Ulery AL, Graham RC, Amrhein C (1993) Wood ash composition and soil pH following intense burning. Soil Sci 156:358–364
Ulrich B (1994) Nutrient and acid-base budgets of central European forest ecosystems. In: Godbold DL, Hüttermann A (eds) Effects of acid rain on forest processes. Wiley, New York, pp 231–264
Ulrich B, Summer ME (1991) Soil acidity. Springer, New York
van Bruggen AHC, Semenov AM (2000) In search of biological indicators for soil health and disease suppression. Appl Soil Ecol 15:13–24
Vance ED (1996) Land application of wood-fired and combination boiler ashes: an overview. J Environ Qual 25:937–944
Verma S, Subehia SK, Sharma SP (2005) Phosphorus fractions in an acid soil continuously fertilized with mineral and organic fertilizers. Biol Fertil Soils 41:295–300
Vieira FCB, Bayer C, Mietniczuk J, Zanatta J, Bissani CA (2008) Long-term acidification of a Brazilian Acrisol as affected by no till cropping systems and nitrogen fertilizer. Aust J Soil Res 46:17–26
Von Uexküll HR, Mutert E (1995) Global extent, development and economic impact of acid soils. In: Date RA, Grundon NJ, Rayment GE, Probert ME (eds) Plant–soil interactions at low pH: principles and management. Kluwer, Dordrecht, pp 5–19
Vorob'eva LA, Avdon'kin AA (2006) Potential soil acidity: notions and parameters. Eurasian Soil Sci 39(4):377–386
Walker TW, Syers JK (1976) The fate of phosphorus during pedogenesis. Geoderma 15(1):1–19
Walter I, Martinez F, Cuevas G (2006) Plant and soil responses to the application of composted MSW in a degraded, semiarid shrubland in central Spain. Compost Sci Util 14:147–154
Wardle DA (2002) Islands as model systems for understanding how species affect ecosystem properties. J Biogeogr 29:583–591
Werkelin J, Skrifvars BJ, Hupa M (2005) Ash-forming elements in four Scandinavian wood species. Part 1: summer harvest. Biomass Bioenergy 29:451–466
Williams CH (1980) Soil acidification under clover pasture. Aust J Exp Agric 20:561–567
Wong MTF, Swift RS (2003) Role of organic matter in alleviating soil acidity. In: Rengel Z (ed) Handbook of soil acidity. Marcel Dekker, New York, pp 337–358
Yunlong C, Smit B (1994) Sustainability in agriculture: a general review. Agric Ecosyst Environ 49:299–307
Zhan G, Erich MS, Ohno T (1996) Release of trace elements from wood ash by nitric acid. Water Air Soil Pollut 88:297–311
Zhang M, Heaney D, Henriquez B, Solberg E, Bittner E (2006) A four-year study on influence of biosolids/MSW cocompost application in less productive soils in Alberta: nutrient dynamics. Compost Sci Util 14:68–80
Zimmermann S, Frey B (2002) Soil respiration and microbial properties in an acid forest soil: effects of wood ash. Soil Biol Biochem 34:1727–1737

Chapter 8
Ash from Combustion of Cacao Residues for Nutrient Recycling: A Case Study

Angelo Sarabèr, Marian Cuperus, and Jan Pels

Abstract A case study has been formulated concerning the use of ashes from combustion of cacao residues (shells) for electricity production and for nutrient recycling to the original soil. The effect in terms of kilograms of fertilizer per hectare and the environmental impact of closing the nutrient and mineral cycle are quantified. If the ashes are used as fertilizer, this fertilizer will only replace about 2% m/m of phosphorus and potassium that is necessary to fulfill the nutrient demand. This means that the contribution of the ashes is small. Furthermore, nitrogen has to be added as fertilizer. There is also a small advantage of reduction of CO_2 emissions by nutrient recycling; this reduction is negligible from the point of view of the plantation, but from the point of view of the filter ash, the potential emission reduction is significant. This study shows that ashes from stand-alone combustion of certain agricultural residues are an potential valuable mineral source for elements such as phosphorus and potassium.

8.1 Introduction

Biomass is one of the sustainable sources of energy that is used for today's production of electricity and heat. Sustainable use of biomass for energy production encompasses many aspects. They range from social aspects such as security of food supply and workers' health to environmental aspects such as clean emission and protection of nature. Although interesting and relevant, this study is limited to only one of those aspects: the role of ash management in nutrient recycling and emission reduction. In Finland and Sweden, for instance, ashes from peat and wood combustion are utilized for fertilization in forestry (Emillson 2006). In 2004, about 27,000 t

A. Sarabèr (✉) and M. Cuperus,
KEMA, P.O. Box 9035, 6800 ET Arnhem, The Netherlands
e-mail: asaraber@vliegasunie.nl

J. Pels
ECN, P.O. Box 1, 1755 ZG Petten, The Netherlands

H. Insam and B.A. Knapp (eds.), *Recycling of Biomass Ashes*,
DOI 10.1007/978-3-642-19354-5_8,

was used for forest fertilization in Finland. Certain wood ashes contain high amounts of calcium compounds, which make them suitable as an alternative liming agent. The calcium carbonate equivalent may be 26–59% depending on the source (Ohno and Erich 1990). Other biomass ashes may also be interesting for fertilization, such as ash from cacao residues as suggested by Simpson et al. (1985), which contain high amounts of potassium.

The Netherlands is one of the biggest producers of cacao products in the world, but all cacao beans are imported from other countries such as Côte d'Ivoire, one of the largest (40%) producers of cacao beans. Some of the process residues (shells) is currently used for direct co-combustion for power generation and some is used in gardening as soil cover. Cacao residues (shells) contain about 8–10% m/m ash-forming matter, which mainly consists of potassium and phosphorous, which are interesting nutrients for agriculture. The caloric value of cacao shells is about 19–22 MJ/kg (higher heating value), whereas the water content is 7–13% m/m (ECN 2010).

An interesting step forward would be to use the ashes from cacao residues as a source for nutrients by recycling the ashes back to the plantations where the cacao was grown or to use them as raw material for fertilizer production. In this case study, the impact on the nutrient balance and reduction of CO_2, NO_x, and SO_2 emissions has been assessed. Disposal of the ashes in a mine is used as a reference.

8.2 Case Description

8.2.1 Methodology and Starting Points

The starting point is that the whole chain has to be analyzed to assess the environmental effects involving closure of the nutrient and mineral cycle. Interviews with representatives from the cacao industry and a literature study were used to define the parameters and to describe the scenarios. The most important part of this chain analysis is the definition of the parameters. The so-called base unit is the most important issue, as it expresses the product under study. The functional unit in this study is fertilization of 1 ha of land in Côte d'Ivoire, on which cacao trees are grown together with shade trees. Further, the effect categories have to be determined, which are the parameters of environmental impact, namely, the nutrient balance and the CO_2, NO_x, and SO_2 emissions.

The starting points for this study are:

- The cacao plantation is situated in Côte d'Ivoire, and the beans are transported to the location of the cacao industry in the Netherlands.
- The cacao shells are a 100% natural residue of the beans and are thermally processed in a stand-alone power plant (so-called bioenergy plant) in the Netherlands.
- The efficiency of the nutrients is comparable to that of the commercial fertilizers which are used on the plantation. Also, the effect of trace elements and

contaminants which are present in the ashes is comparable to that of the commercial fertilizers.

The chemical composition of the ashes is predicted with a modified version of the KEMA Trace Model® (KEMA 2005). The model is an empirical and statistical computer model to predict emissions and ash composition of a dry pulverized coal-fired boiler including biomass co-combustion. A modified version of the model has been developed for fluidized bed boilers, taking into account the lower combustion temperature, the use of inert or reactive bed material, and the distribution of ash and bed material.

8.2.2 *Two Scenarios*

Two scenarios have been examined and compared in this study. These scenarios are:

- Scenario 1 (Fig. 8.1): The cacao shells are combusted in a bioenergy plant in the Netherlands. The ash produced is transported to a sea port in the Netherlands. From this port the ash is shipped to Côte d'Ivoire. The ash is transported by truck to a cacao plantation. The ash is used as fertilizer. The ash contains no nitrogen and not enough phosphorous and potassium. Therefore, extra NPK-containing fertilizer is used to fulfill the nutrient demand. This NPK fertilizer is produced in Côte d'Ivoire.
- Scenario 2 (Fig. 8.2): The cacao shells are combusted in a bioenergy plant in the Netherlands. The ash produced is transported to a mine in Germany as filling material. The fertilization of the cacao plantation is performed by means of an artificial fertilizer (NPK). The fertilizer is produced in Côte d'Ivoire.

The system has to be bounded for the analysis. The boundaries determine the scope of the study. There are four subsystems in this study of importance:

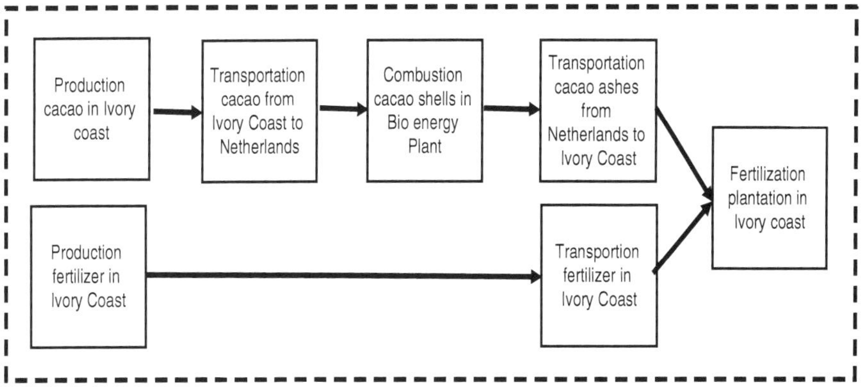

Fig. 8.1 System boundaries of scenario 1

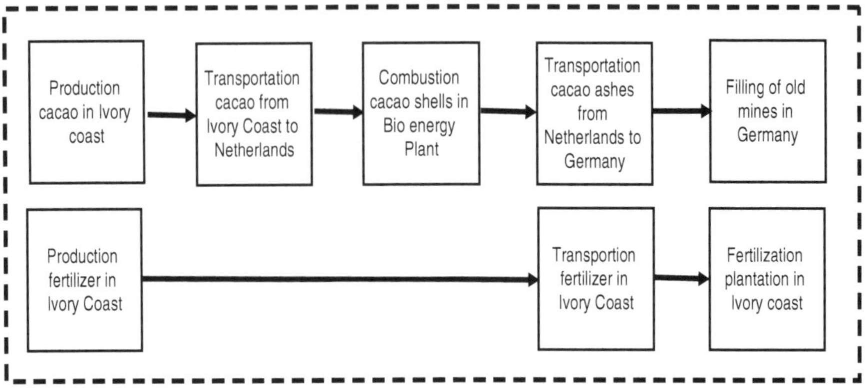

Fig. 8.2 System boundaries of scenario 2

1. The cacao plantation subsystem
2. The cacao shell combustion subsystem
3. The cacao shell ash subsystem
4. The fertilizer production subsystem

8.2.3 Cacao Cultivation

Cacao (*Theobroma cacao*) is a small evergreen tree in the family *Sterculiaceae* or *Malvaceae*. Cacao is grown in more than 30 countries around the world in Africa, South America, and Asia, principally in areas that fall within 20°N and 20°S of the equator (FAO 2009). The cacao tree is an understory tree, growing best with some overhead shade. The seeds of the cacao tree are used to make cacao and chocolate. The fruit, called a cacao pod, contains 20–60 seeds, usually called beans. The pods consist of the husks and the beans. Every bean is surrounded by a thin shell. The pods are harvested, the husks are removed, and the beans are dried and fermented. After fermentation, drying, and packing, the beans, including the shells, are transported to the location of the cacao industry.

8.2.4 Combustion of Cacao Shells for Heat and Power Generation

In this case it is assumed that the beans are transported to the Netherlands. The beans are stored in the harbor before transport to the cacao processing plant. Beans are broken to nib and the shells are separated. The nib is processed to cacao products. The shells are combusted or used as gardening material. All materials are transported by road or water. An example of cacao shells is given in Fig. 8.3.

For the scenarios it is assumed that the cacao shells are combusted in a circulating fluidized bed combustion plant to produce heat and power. This is a type of boiler for production of energy from biomass on a large scale (so-called bioenergy

Fig. 8.3 Cacao shells: a residue of processing cacao beans

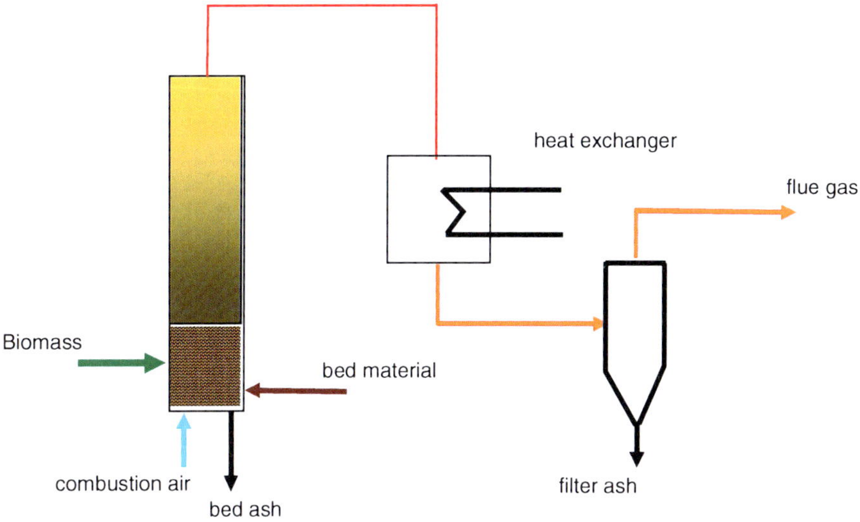

Fig. 8.4 Basic design of the circulating fluidized bed combustion plant

plants) (Fig. 8.4). The general data of the bioenergy plant that were used in this study are given in Table 8.1. Sand is used as bed material. Some limestone is used to capture sufficient SO_2 to meet the Dutch emission requirements.

The predicted chemical composition of the filter ash is presented in Table 8.2. The macroelements are also expressed as oxides (this does not mean that these elements

Table 8.1 General data of the bioenergy plant

Parameter	Unit	Value
Net power	MW_e	30
Net electrical efficiency	%	29.0
Full load hours	h	6,000
Net electrical capacity	MW_e	29
Use of cacao shells (as received)	t/h	23
Use of bed material	kg/h	824
Percentage of ash that is filter ash	%	80
Percentage of ash that is bed ash	%	20

Table 8.2 Predicted composition of filter ash from the bioenergy plant

Macroelements (%)			
Based on elements		Based on oxides	
Al	0.6	Al_2O_3	1.2
Ca	13.6	CaO	19
Cl	0.2		
Fe	2.8	Fe_2O_3	4.0
K	39	K_2O	47
Mg	2.9	MgO	4.8
Na	0.4	Na_2O	0.5
P	4.2	P_2O_5	10
Si	4.3	SiO_2	9
Ti	0.3	TiO_2	0.5
Residual fuel			5
Microelements (mg/kg)			
As	3.3	Mo	1.1
Ba	285	Ni	143
Be	0.2	Pb	10
Cd	1.0	Sb	26.9
Co	45	Se	0.15
Cr	78	Sn	40.8
Cu	514	Te	1.0
F	3.2	U	1.6
Hg	0.10	V	25
Mn	946	Zn	985

are present as oxides). The main nutrients in filter ash are potassium, calcium, and phosphorus. Thermodynamic calculations were performed to predict the potential compounds in cacao ash under equilibrium conditions. These were performed using FactSage©. The main compounds will be K_2SO_4, K_2CO_3, and $Ca_5HO_{13}P_3$. It is assumed that the filter ash contains 5% residual pyrolized cacao shells (carbon).

8.2.5 Basic Assumptions of the Two Scenarios

In Côte d'Ivoire the production of dried beans in 2005 was 740 kg/ha (Elzebroek and Wind 2008). According to ADM Cacao (2007 Interview with F. de Kort ADM

Table 8.3 Emission per ton kilometer of transportation (Eclipse 2003)

		Road: heavy lorry trailer (40 t)	Medium-sized lorry (regional transport)	Sea: bulk carrier
CO_2	(g/t km)	60.1	156	7.97
NO_x	(g/t km)	0.99	2.56	0.21
SO_2	(g/t km)	0.045	0.115	0.20

Cacao 22 November 2007. Koog aan de Zaan, the Netherlands), 10% of the beans consist of a shell, which means that per hectare 74 kg cacao shells is produced.

For the environmental impact of transportation many data are available. In this study the Eclipse data were used. Eclipse (*E*nvironmental and E*c*ological *L*ife Cycle *I*nventories for Present and Future *P*ower *S*ystems in *E*urope) was funded by the EU and was carried out in 2002–2003. One of the objectives of Eclipse was to provide a harmonized set of public, coherent, transparent, and updated data on new and decentralized power systems for life cycle analyses.

The transportation by road in Europe is carried out by means of a heavy lorry trailer (40 t). The transportation by sea is carried out by means of a bulk carrier. The transportation by road in Côte d'Ivoire is carried out with a medium lorry. In Table 8.3, an overview is given of the emissions per ton kilometer of these vehicles. An overview of the assumed distances used in this study is given in Table 8.10. Transport of cacao beans to the Netherlands and transport of shells to bioenergy plants are not included, as these are present in both scenarios.

8.3 Mineral Balance

To estimate the amount of nutrients needed for fertilization of the soil, an estimate of the nutrient balance is needed. The plant density and the climate are important (IFA 2008). Nutrient stocks have been restricted to the upper 30 cm, as most feeding roots of cacao are concentrated at that depth. Removal of nutrients from cacao ecosystems is caused by yield (beans and husks), immobilization in stem and branches, and leaching of nutrients below the rooting zone (Hartemink 2005). Most nutrients in cacao ecosystems are lost by the harvest of beans and husks. In Table 8.4, an overview is given of the nutrient removal caused by the crop of 1 ha (740 kg dry cacao beans and 1.0 t cacao husks).

For the nutrient demand there are several recommendations for cacao. The recommendations for the nutrient demand for 1 ha of cacao plantation differ hugely, as shown in Table 8.5. We decided to use the statistical approach: all improbable data have been removed.

For the mineral balance it is assumed that the husks are returned to the soil and that all the available nutrients are reused. This assumption is assessed in the sensitivity analysis. In Table 8.6, an overview is given of the mineral balance. The husks are supposed to be returned to the soil, so these nutrients are directly

Table 8.4 Nutrient removal in kilograms caused by the crop of 1 ha of plantation in Côte d'Ivoire (IFA 2008)

Beans (+shells)					Husks					Total				
N	P_2O_5	K_2O	MgO	CaO	N	P_2O_5	K_2O	MgO	CaO	N	P_2O_5	K_2O	MgO	CaO
16.4	5.0	6.7	1.7	0.4	9.8	3.1	38.4	6.9	3.1	26.1	8.1	45.1	8.6	3.6

Table 8.5 Fertility recommendations according to different literature sources (kg/ha)

References	N	P	K
Elzebroek and Wind (2008)	50–100	25	75
FAO (2009)	0	28	32
CABI (2009)[a]	200	25	300
Uribe et al. (2001)	100	90	200
IFA (2008)	147	8	106
CPCRI (2009)	110	10	64
Average	126	26	202

[a]Before pod production

Table 8.6 Mineral balance for a cacao plantation in Côte d'Ivoire (kg/ha)

	Nutrient demand	Addition by husks	Natural addition of nutrients	Amount to be added by means of fertilizer/ash
N	126	13.2	43.3	69.5
P	26	1.83	0.30	23.9
K	202	43.1	95	63.9

Table 8.7 Emission data (kg) for triple superphosphate and potassium chloride (source Simapro 7.1.8 using Ecoinvent 2.0 database)

Emissions	Per kilogram P_2O_5	Per kilogram K_2O
CO_2	2.016	0.484
CO_2eq	2.064	0.533
NO_x	0.0072	0.0016
SO_2	0.028	0.0083

recycled to the plantation. The nutrient demand minus the natural addition of nutrients minus the husks gives the amount to be added by fertilization. Further, some natural addition occurs by deposition and transfer. Data from plantations in Cameroon (Hartemink 2005) are used, as data for Côte d'Ivoire were not found in the literature.

The preferred nutrient sources for the nitrogen, phosphorous, and potassium fertilizers for a cacao plantation are urea, triple superphosphate (with 48% P_2O_5), and potassium chloride (with 49% K_2O), respectively (IFA 2008). It is assumed that the fertilizer is produced in Côte d'Ivoire. Data for production in Europe were used, as other data were not available (Table 8.7).

8.4 Environmental Impact Analysis

The environmental impact analysis is focused on emissions, especially those of CO_2, NO_x, and SO_2. The emissions related to ash transport and fertilizer production and transport need to be calculated to obtain the total environmental impact.

8.4.1 Combustion of the Cacao Shells in the Bioenergy Plant

Per hectare cacao plantation in Côte d'Ivoire, 74 kg cacao shells are produced. These shells are combusted in a bioenergy plant. In Table 8.8, an overview is given of the main characteristics of the circulating fluidized bed combustion plant for the combustion of 74 kg cacao shells. The combustion of 74 cacao shells produces 5.3 kg filter ashes. Only the filter ash is suitable as a fertilizer. The bed ash consists of approximately 69% bed material (sand) and additive, whereas the filter ash contains only 13% bed material and additive. It is assumed that all potassium is present as K_2CO_3.

8.4.2 Amount of NPK Fertilizer Needed

In Table 8.9, an overview of the amounts of fertilizer/ash to be used on 1-ha plantations is given for both scenarios. The amount of fertilizer in scenario 2 needed to compensate for the amount of filter ash used in scenario 1 is 1.1 kg/ha for triple superphosphate and 4.1 kg/ha for potassium chloride.

Table 8.8 Overview of the main characteristics of the combustion of 74 kg cacao shells in the bioenergy plant

Characteristics	Amount (kg)
Amount of fuel used (as received)	74
Amount of bed material used	2.7
Amount of additive used (lime)	1.0
Amount of filter ash produced, including bed material and additive	6.7

Table 8.9 Overview of the amounts of fertilizer/ash needed for the two scenarios (kg/ha)

	Demand	Scenario 1		Scenario 2
		Added by filter ash	Synthetic fertilizer	Synthetic fertilizer
N	69.5	0	298 kg urea	298 kg urea
P	23.9	0.22	113 kg triple superphosphate	114 kg triple superphosphate
K	63.9	2.05	125 kg potassium chloride	129 kg potassium chloride

8.4.3 Overview of the Process Steps for the Two Scenarios

As already stated, the functional unit in this study is the fertilization of 1 ha of land in Côte d'Ivoire, on which cacao trees are grown together with shadow trees. In Table 8.10, an overview is given of the process steps for the two scenarios. As can be seen from Table 8.10, several process steps are identical for both scenarios. This means that these process steps are *not* taken into account.

8.4.4 Scenario 1: Recycling of the Ashes as a Fertilizer

The filter ash is transported by means of a heavy lorry trailer from the bioenergy plant to the port. The distance between these is assumed to be 100 km. Table 8.3 gives the emissions per ton kilometer. Per hectare cacao plantation in Côte d'Ivoire, 74 kg cacao shells are produced, resulting in 6.7 kg filter ash, corresponding to 0.67 t km. The filter ash is transported by means of a bulk carrier from the Dutch port to Côte d'Ivoire. The distance is assumed to be 6,000 km (SenterNovem 2007), which results in 40.2 t km. The ash is transported by means of a medium-sized lorry from the port in Côte d'Ivoire to the cacao plantation. The distance is assumed to be 800 km, which results in 5.4 t km.

Table 8.10 Overview of the process steps for PK fertilization of 1 ha cacao plantation in Côte d'Ivoire (N fertilization excluded)

Process step	Scenario 1	Scenario 2
Production of cacao beans in Côte d'Ivoire	740 kg dried beans	
Transportation of cacao beans from Côte d'Ivoire to the Netherlands	740 kg fried beans 6,000 km	
Production of cacao shells	74 kg cacao shells	
Transportation of cacao shells to bioenergy plant	74 kg cacao shells 100 km	
Combustion in bioenergy plant	74 kg cacao shells 6.7 kg filter ash	
Transportation of ash to port in the Netherlands	6.7 kg filter ash 100 km	–
Transportation of ash to salt mine in Germany	–	6.7 kg filter ash 250 km
Transportation of ash to Côte d'Ivoire	6.7 kg filter ash 6,000 km	–
Transportation of ash to cacao plantation	6.7 kg filter ash 800 km	–
Production of fertilizer in Côte d'Ivoire	113 kg triple superphosphate 125 kg potassium chloride	114 kg triple superphosphate 129 kg potassium chloride
Transportation of fertilizer in Côte d'Ivoire	238 kg PK fertilizer 800 km	243 kg PK fertilizer 800 km

Table 8.11 Overview of the CO_2, NO_x, and SO_2 emissions caused by scenarios 1 and 2 for PK fertilization of 1 ha

	CO_2 (kg)	NO_x (g)	SO_2 (g)
Scenario 1			
Transportation of ash from bioenergy plant to Dutch port	0.04	0.66	0.03
Transportation of ash from Dutch port to Côte d'Ivoire	0.32	8.4	8.0
Transportation of ash from port in Côte d'Ivoire to cacao plantation	0.84	14	0.62
Production of PK fertilizer	145	487	2,004
Transportation of fertilizer to cacao plantation	30	487	22
Total	175.5	997	2,034
Scenario 2			
Transportation of ash from circulating fluidized bed combustion plant to German mine	0.10	1.7	0.08
Production of PK fertilizer	147	493	2,033
Transportation of fertilizer to cacao plantation	30	498	22
Total	177.1	993	2,056
Difference (scenario 2 minus scenario 1)	1.6	−4.4	21

The major part of the PK nutrient demand has to be provided by synthetic fertilizer. Table 8.7 shows the emissions produced by 1 kg of these fertilizers. These emissions are valid for production of the fertilizer in Europe. We assume the production of the fertilizers takes place in Côte d'Ivoire. However, because of lack of better data, the production data for Europe are used. The fertilizer (238 kg) is transported within Côte d'Ivoire by means of a medium-sized lorry. The distance is assumed to be 800 km, corresponding to 190 t km. In Table 8.11, an overview is given of the total emissions in scenario 1.

8.4.5 Scenario 2: Use of an Artificial Fertilizer

The filter ash is transported by means of a heavy lorry trailer from the bioenergy plant to a mine in Germany. The distance is assumed to be 250 km. The amount of ash transported is 6.7 kg, which corresponds to 1.68 t km. The total PK nutrient demand has to be provided by synthetic fertilizer (243 kg) under the same conditions as described in scenario 1. The transport distance corresponds to 194 t km. In Table 8.11, an overview is given of the total emissions in scenario 2.

8.4.6 Comparison of the Two Scenarios

In Table 8.11, the emissions caused by the two scenarios are compared. The emissions caused by scenario 1 and by scenario 2 are about the same, which is not surprising because of the low level of replacement of the PK fertilizer that can be obtained. However, the potential CO_2 reduction per kilogram of filter ash is significant, namely, about 0.40 kg CO_2/kg filter ash.

In this study some assumptions have been made. In the sensitivity analysis, the influence of these assumptions on the emissions is determined.

- All of the phosphorous and potassium present in the ash is available as nutrient. If the availability of both elements drops below 40–45%, then the CO_2 and SO_2 emissions in scenario 1 will be higher than those in scenario 2.
- The transport distance within Côte d'Ivoire is 800 km. If the transport distance is 500 km, the CO_2 emissions in scenarios 1 and 2 decrease to 160.3 and 161.9 kg, respectively. If the transport distance increases to 1,000 km, the CO_2 emissions are 179.3 and 180.9 kg, respectively. This means that the absolute value in both scenarios will be influenced to a small extent, but the difference between the values in both scenarios remains the same.
- The transport distance within Côte d'Ivoire is the same for the filter ash and the PK fertilizer. If the transport distance of the ash is about 1,500 km more than that for the PK fertilizer, then the CO_2 emissions in scenario 1 will be higher than those in scenario 2.

8.5 Discussion

From an environmental point of view, the impact of recycling the ashes originating from the yield of 1 ha of cacao beans are very small. The ratio of phosphorous to potassium that is present in the ash is more or less comparable with the ratio that is present in the artificial fertilizer. Therefore, it is possible to replace only a small part of the PK fertilizer that is needed for the fertilization of 1 ha. If the ashes are used as fertilizer, not more than 5.2 kg synthetic PK fertilizer will be replaced per hectare. However, to fertilize the soil properly, another 243 kg synthetic PK fertilizer (and about 298 kg urea) per hectare is necessary (depending on the nutrient need). So only a very small part (about 2%) of the synthetic PK fertilizer can be replaced, assuming that the uptake by the cacao tree is the same as for commercial fertilizer.

It is important to examine how much of the phosphorous and potassium present in the ash will be effectively available as nutrient. The nutrients in artificial fertilizer are available instantly. Thermodynamic calculations showed that the most probable potassium compounds are sulfates and carbonates, which are also highly soluble compounds, but that phosphorous may be present as a less soluble compound. No research has been carried out yet on the nutrient demand of the soil for trace elements. These trace elements (Fe, Zn, Cu, Mo, B, and Mn) are present in both the synthetic fertilizer and the filter ash. Also, the influence of potentially toxic elements (Pb, Cd, Cr^{VI}, Hg) on the fertilization of the soil and on the health of the cacao trees has not been examined. In a recent study on the suitability of alleviating low pH, Bougnom et al. (2009, 2010, 2011) have shown that composts prepared with ashes could be suitable for remediating tropical soils.

In this study only the emissions of CO_2, NO_x, and SO_2 in both scenarios have been compared. The replacement of artificial fertilizer with the ashes diminishes the amount of CO_2 SO_2 emitted per hectare by 1.6 kg and 21 g, respectively. This is negligible from the point of view of the plantation, but from the point of view of the filter ash, the potential of emission reduction is attractive. This makes the ash a potentially valuable mineral source.

In this case study it is assumed that the cacao residues are combusted in the Netherlands and have to be transported over a large distance to Côte d'Ivoire. This can be avoided by using the ash as a raw material in fertilizer production in the Netherlands itself. This will reduce the overall sea transport of ash to Côte d'Ivoire and raw materials for fertilizer production to the Netherlands.

8.6 Conclusions

A case study has been performed to assess the environmental impact and feasibility of ash recycling to restore the nutrient balance of 1 ha of a sustainable plantation in Côte d'Ivoire. From a limited environmental analysis, in which only the emissions of CO_2, NO_x, and SO_2 were compared, it can be concluded that:

If the ashes generated by combustion of cacao residues (shells) derived from cacao beans are used as fertilizer on a cacao plantation, this fertilizer will replace only about 2% m/m of phosphorus and potassium that is necessary to fulfill the recommended fertilization. This means that the contribution of the ashes is negligible. Furthermore, nitrogen has to be added as fertilizer because the ashes do not contain nitrogen

The replacement of artificial fertilizer with ashes diminishes the amount of CO_2 and SO_2 emitted per hectare by about 1.6 kg and 21 g, respectively, despite the reverse logistics. However, the NO_x emissions increase by 4 g. This is negligible from the point of view of the plantation, but the potential emission reduction per kilogram of filter ash is significant.

It will be more interesting to use the ashes generated as raw material for fertilizer production in the region where the ashes are generated (i.e., the Netherlands). The accompanying CO_2, NO_x, and SO_2 emissions due to transport will be reduced. Another advantage is the less complex logistics.

This study shows that ashes from stand-alone combustion of certain agricultural residues are a potentially valuable mineral source for elements such as phosphorus and potassium.

References

ADM Cacao (2007) Interview with F. de Kort ADM Cacao 22 November 2007. Koog aan de Zaan, the Netherlands

Bougnom BP, Maier J, Etoa FX, Insam H (2009) Composts with wood ash addition: a risk or a chance for ameliorating acid tropical soils? Geoderma 153:402–407

Bougnom BP, Knapp BA, Etoa FX, Elhottová D, Němcová A, Insam H (2010) Designer composts for ameliorating acid tropical soils: effects on the soil microbiota. Appl Soil Ecol 45:319–324

Bougnom BP, Knapp BA, Etoa FX, Insam H (2011) Possible use of wood ash and compost for improving acid tropical soils. In: Insam H, Knapp BA (eds) Recycling of biomass ashes. Springer, Heidelberg, pp 87–105

CABI (2009) http://www.cabi-commodities.org. Accessed Jan 2010

CPCRI (2009) Central Plantation Crops Research Institute. Garden management – Cocoa. www.cpcri.gov.in. accessed Jun 2009

Eclipse (2003) Environmental and ecological life cycle inventories for present and future power systems in Europe. Biomass. Workpackage 3. More information on the project is available at http://www.eclipse-eu.org. Accessed 20 Jan 2008

ECN (2010) Phyllis database. Available via http://www.ecn.nl. Accessed 26 Apr 2010

Elzebroek ATG, Wind K (2008) Guide to cultivated plants. Kyodo, Singapore

Emillson S (2006) From extraction of forest fuels to ash recycling. International handbook. Published by Swedish Forest Agency

FAO (2009) http://ecocrop.fao.org/ecocrop/srv/en/dataSheet?id=2074. Accessed 28 Jun 2009

Hartemink AE (2005) Nutrient stocks, nutrient cycling and soil changes in cacao ecosystems – a review. Adv Agron 86:227–253

IFA (2008) World fertilizer use manual, Cacao. J. Snoeck and P. Jadin, Institut de Recherches du Café, du Cacao et Autres Plantes Stimulantes (IRCC/CIRAD), Montpellier. http://www.fertilizer.org/ifa/publicat/html/pubman/cacao.htm. Accessed 15 Jan 2008

KEMA (2005) The fate of trace elements during co-combustion of biomass in coal-fired power plants. Paper presented at the Clearwater Coal conference, 2005

Ohno T, Erich MS (1990) Effect of wood ash application on soil pH and soil test levels. Agric Ecosyst Environ 32:433–438

SenterNovem (2007) (B. Bruinsma) Uitgebreide energie studie voor de cacao-industrie. Rapportnummer 2605201DR01. Utrecht

Simpson BK, Oldham JH, Martin AM (1985) Extraction of potash from cacao pod husks. Agric Wastes 13:69–73

Uribe A, Mendez H, Mantilla J (2001) Effect of balanced fertilization on cocoa yield in Colombia. Available via http://www.ipni.net. Accessed 25 Apr 2010

Chapter 9
The Use of Biomass Combustion Ashes in Brick Making

Fritz Moedinger

Abstract The addition of biomass ashes to a brick feed has been investigated in a full-scale industrial production process over several days. The addition of biomass ashes to the brick feed is part of an ongoing research and development project targeted at substituting or combining quarried raw materials with suitable waste materials. In this chapter some of the early results of the first industrial trial runs with biomass ashes are presented. The main scope of this ongoing research and development project is reduction of production cost, generation of additional revenue from gate fees, and improvement of product characteristics.

9.1 Introduction

Besides other methods of recycling biomass ashes as described in other chapters in this book (see Chap. 1, Knapp and Insam 2011; Chap. 6, Omil et al. 2011; Chap. 11, Ribbing and Bjurström 2011), another option is to use the ashes for producing construction materials such as concrete (see Chap. 10, Berra et al. 2011) and brick.

For the production of ceramic bricks, the predominant raw material used is mineral clay. Any good brick clay should have low shrinkage and low swelling characteristics, consistent firing color, and a relatively low firing temperature, but at the same time produce an adequately dry and fire-strength brick. The guiding rule of choice on wastes and by-products must rest on their compatibility with the original (host) raw material being used, whereas they must not degrade the final product by focusing simply on making it a repository for wastes. Thus, it is necessary to

F. Moedinger
University of Staffordshire/Recuperi Industriali S.r.l. Fritz Moedinger, Via Don Bosco 10, 39042 Bressanone, Brixen, Italy
e-mail: fritz.moedinger@gmail.com

H. Insam and B.A. Knapp (eds.), *Recycling of Biomass Ashes*,
DOI 10.1007/978-3-642-19354-5_9,

establish a company-and-production- site-specific "tailor-made" quality product standard. In general, the firing of a mineral mass rich in aluminum silicate such as brick clay allows permanent stabilization of almost all heavy metals included except chromium (Anderson and Skerratt 2003).

9.2 Clay as a Raw Material in Brick Making

In compositional terms, good brick clay should feature:

- Grain size distribution (sieve line): A high percentage of larger particles requires higher sintering temperatures, resulting in a greater energy requirement and longer firing times.
- Accessory minerals: Quartz, feldspar, and amphibole have an effect on the sintering behavior and might result in undesired colors.
- Organics: Create voids in the finished product. If the organics have a high sulfur content, this might have effects on color, the fumes, and the kiln atmosphere.
- Sulfur: Pyrite and marcasite release SO_x on firing, creating large-diameter pores in the brick that might reduce compressive and flexural strength.
- Carbonate minerals: Calcite, dolomite, and other carbonate minerals, such as ankerite and siderite, do have, if finely dispersed, an effect on the release of low-temperature-carbonization gases owing to the formation of channels and funnels but in larger quantities will reduce the compressive and flexural strength of the final product and might lead to chipping on the surface.
- Alkalis: A low alkaline earth content, magnesium and calcium being the most common, is desirable to avoid firing interactions which could promote discoloration of the final product.
- Metal oxides and hydroxides: Goethite and hematite, for example, both containing iron oxide (the chief colorant responsible), which ideally should be in the range 5–12% (for good strong color), are the main origin of the red brick color.
- Natural radioactivity of the raw materials (radon).

Adding foreign substances to influence the physical properties of the finished product or the workability is nothing new. It is important to distinguish between "additions" to the clay body (such as saw dust or paper sludge), and "substitutions" (such as sewage sludge, ash, or treated aluminum salt slag) that replace a part of the original clay body. The distinction between "addition" and "substitution" for extraneous materials is not always straightforward or easy: Any substance that is added to the original clay body without substitution of clay modifying its inherent characteristics might be considered an addition. A substitution, on the other hand, may be viewed as any material that for a required volume of brick reduces the quantity of clay needed to achieve that specific volume target. A substitution can also modify the clay body.

9.3 Brick Production Process

The brick production process in itself has remained unchanged for a very long time: a shapeable mass is formed into a brick, assumed to be a small, regularly shaped unit that a bricklayer can grasp with one hand while picking up mortar with the other. A typical modern production process is shown in Fig. 9.1.

Additions to the brick feed can take place either at the beginning of the process or in later steps.

In the trial runs explained later, biomass ashes were introduced as a substitution displacing quarried raw materials. They were mixed with the other raw materials right at the beginning of the process.

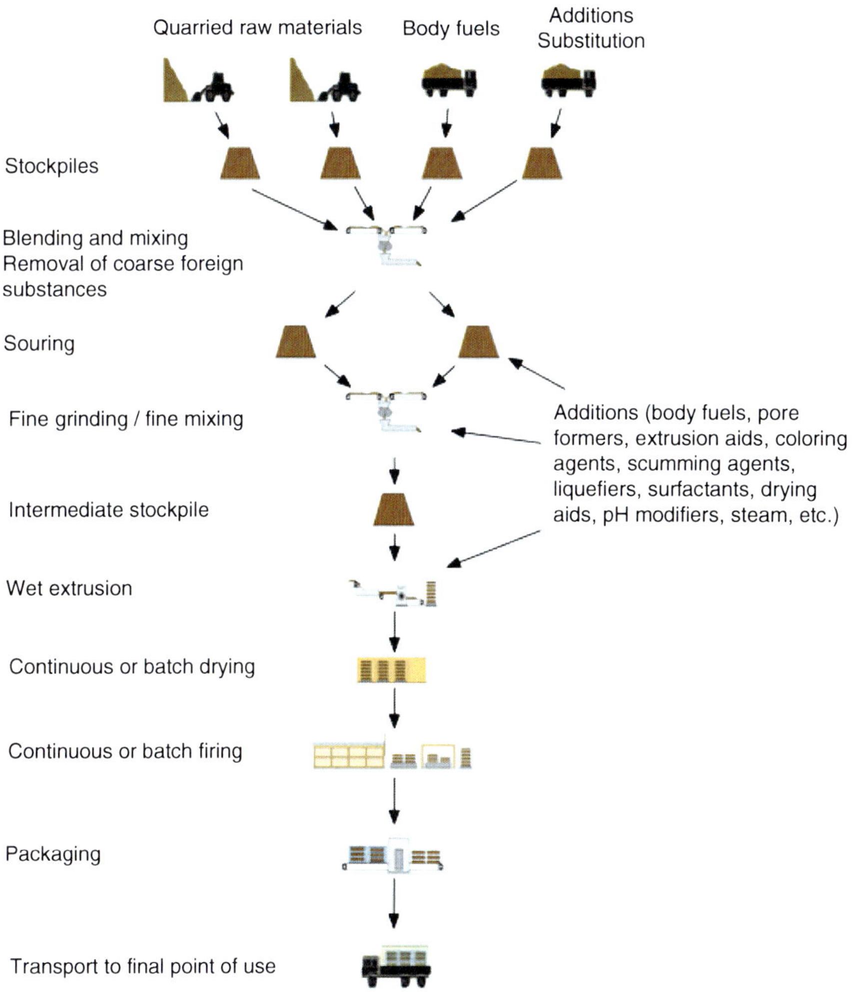

Fig. 9.1 Brick manufacturing process

9.4 Use of Waste-Based Additions or Substitutions to Brick Feeds

In most cases waste disposal laws and regulations require that the waste to be disposed of be accompanied by some kind of analytical data. However, such data are generally insufficient to determine whether the waste can be used in brick. For a typical brick factory the information that is required is listed in Table 9.1 (Moedinger and D'Anna 2002; Moedinger 2003, 2004).

It is essential that the quality and composition of the waste in use at the brick plant be continuously monitored so one is aware of any sporadic fluctuations in composition that could detrimentally affect the manufactured product. Long-term

Table 9.1 Basic information for testing the potential waste material for inclusion in a brick body

Oxides and heavy metals	Other	Mineralogy	Impacts
Aluminum oxide (Al_2O_3)	Humidity (H_2O%)	Kaolinite (%)	Odor
Sulfur trioxide (SO_3)	Acidity (pH)	Smectite (%)	Odor (on drying)
Antimony oxide (Sb_2O_3)	Conductivity (μS/cm)	Illite (%)	Firing color
Barium oxides (BaO)	Dry substance (ç)	Muscovite (%)	Shrinkage (drying – firing)
Boron trioxide (B_2O_3)	Not combusted (%)	Chloride (%)	Cracking (drying – firing)
Calcium Oxide (CaO)	Chloride (%)	Mica (%)	Evaluation of possible emissions
Iron oxide (Fe_2O_3)	Sulfur (%)	Fluor (%)	Evaluation of possible health risks
Phosphorus pentoxide (P_2O_5)	Fineness (percentage clay minerals if any)		Evaluation of possible financial benefits
Magnesium oxide (MgO)	Plasticity		Evaluation of possible liability risks
Manganese oxide (MnO)	Firing temperature		Leaching on firing
Lead oxide (PbO)	Mechanical properties at different firing temperatures		
Potassium oxide (K_2O)			
Silica (SiO_2)			
Sodium oxide (Na_2O)			
Titanium dioxide (TiO_2)			
Zinc oxide (ZnO)			
Zirconium oxide (ZrO_2)			
Arsenic (As)			
Cadmium (Cd)			
Chromium (Cr)			
Mercury (Hg)			
Nickel (Ni)			
Copper (Cu)			
Selenium (Se)			
Lead (Pb)			
Zinc (Zn)			
Vanadium (V)			

production tests are necessary to establish eventual variations in the composition and the impact on the product or production process.

Some of the potentially detrimental results on brick products or the production process caused by various wastes can be offset by the use of appropriate "corrective" additives:

- The concentration of chromium and chlorates with respect to their possible volatilization on firing and their subsequent concentration in the flue gases and potential effects on the refractory material of the kiln
- Odor and smells
- Heavy metals
- Organic contamination
- Chemical contamination
- Particle size distribution
- Water absorption
- Content of carbonate minerals
- Soluble salts

9.5 Body Fuels

Body fuels, i.e., combustible substances added to the brick feed that are then combusted upon firing the brick, have been in use since the Egyptians added straw to the brick feed. Owing to the remaining organic content, biomass combustion ashes might be considered a low calorific value body fuel.

A tunnel kiln (Fig. 9.2) works as a counterflow heat exchanger. In the tunnel kiln packs of bricks set on a car train on rails move through the kiln one after the other. During their journey, the cars move toward, through, and past the stationary firing section at the center of the structure. During its travel, the brick set on the kiln car is slowly, ideally uniformly, heated up to the required firing temperature and then cooled down again (Fig. 9.3). Heat transfer within a tunnel kiln and within the

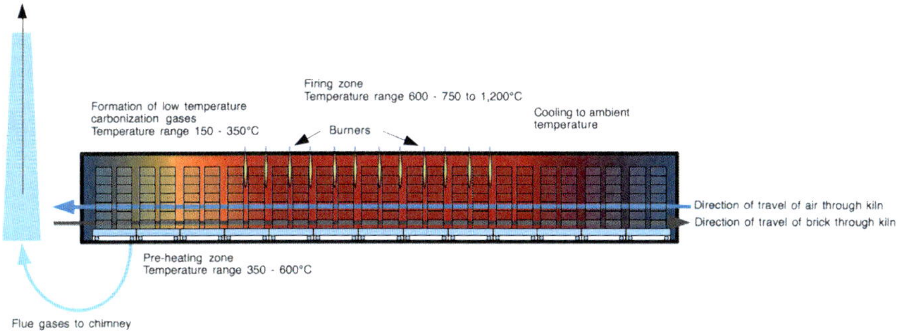

Fig. 9.2 Tunnel kiln

usually densely packed bricks on the kiln car takes place primarily by forced convection of the turbulent kiln gas flow, and, limited to the firing zone, by radiance of the burner flame to the brick. Further heat transfer by conduction from brick to brick occurs at the contact surfaces of the tightly stacked bricks. At a temperature range of 150–350°C the volatile proportion of any organic addition made to the brick feed is released as low-temperature-carbonization gases. These gases are usually conveyed to the chimney or to an appropriate postcombustion system.

As heating of the bricks set on the kiln car does not take place uniformly across the section of the brick, release of low-temperature-combustion gases at different points of the travel through the kiln can cause problems. The firing curve pictured in Fig. 9.4 shows the delayed reaction of the energy rich additions to the clay body:

Fig. 9.3 Methods of stacking bricks on kiln cars

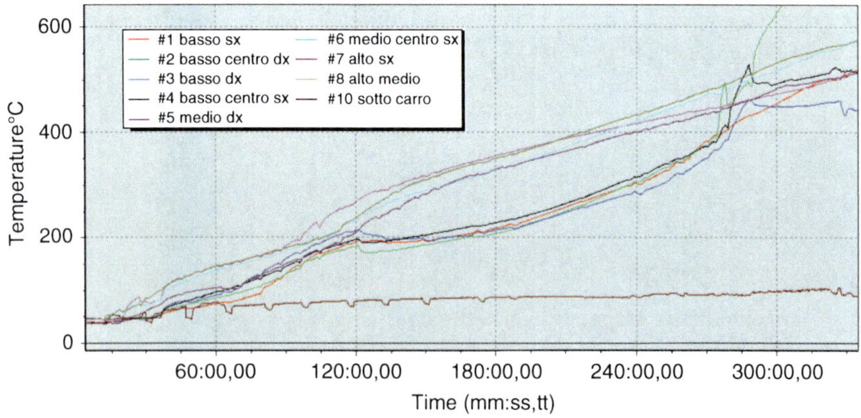

Fig. 9.4 Example of release of energy of combustible organics in a tunnel kiln (Legend: #1 basso sx -lower left side; #2 basso centro dx - lower center right side; #3 basso dx - lower right side; #4 basso centro sx - lower center left side; #5 medio dx - middle right side; #6 medio centro sx - middle center left side; #7 alto sx - top left side; #8 alto medio - top center; #10 sotto carro - below the cart)

The combustion of the energy content in the core of the pile occurs later than that on the perimeter.

Only at a temperature of about 550–600°C do the temperature differences between the lower and the upper layers of the brick stacked on the kiln car diminish considerably.

The problems of using body fuels featuring a high percentage of volatiles is explained here with the example of paper sludge: At a temperature of about 150°C, all water that is still present in the cellulose is evaporated. This evaporation process is concurrent to driving out water still present in the clay body and extends into the first phases of release of crystalline water. At temperatures between 100 and 200°C, volatile substances are dissociated and evaporate, releasing carbon monoxide, hydrogen, and hydrocarbons.

Recent research by the author has shown the effects of concurrent combustion (a substance with an otherwise higher ignition temperature is ignited by the combustion of a substance with a lower ignition temperature) and the impact on the energy balance of a tunnel kiln. Such an example is pictured in Fig. 9.5 for a brick feed with 20% of paper sludge added to the clay in addition to a 1.5% of a bituminous coal featuring a high volatile content.

The above-mentioned differential thermal analysis (DTA) is of particular interest when it is compared with the DTA for the same basic mix but with 1.5% of anthracite and less than 2.5% of volatiles instead of the bituminous coal in the previous example. The DTA in Fig. 9.6 shows release of energy at the same temperatures as in the previous case but a second, smaller peak is observed at higher temperatures as well. This second peak has a positive impact on the thermal balance of the kiln.

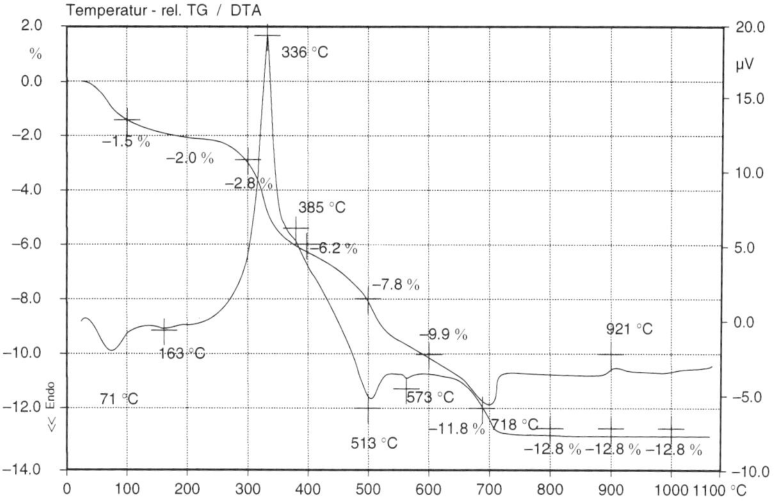

Fig. 9.5 Differential thermal analysis (DTA) of brick feed with 20% by volume paper sludge and 1.5% by volume of a bituminous coal

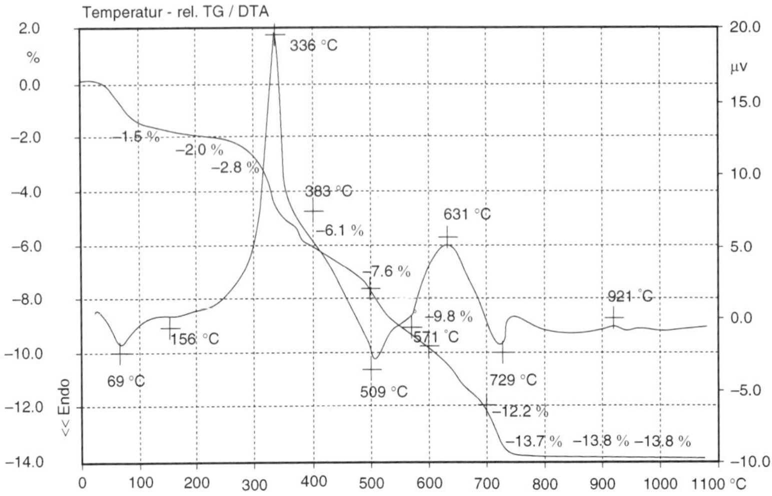

Fig. 9.6 Differential thermal analysis (DTA) of brick feed with 20% by volume paper sludge, 3% of hydrophobized wood combustion ash, and 1.5% of anthracite

An effect similar to that observed for the anthracite/paper sludge mix was observed in the tunnel kiln when a biomass ash/paper sludge mix was fired. This effect has to be confirmed in further and longer-lasting tests to prove that, biomass combustion ashes contribute positively to the energy balance of a brick tunnel kiln.

9.6 Industrial Testing

The data were obtained in industrial testing with manufacturing batches of about 1,750 t of brick. The daily production capacity of the plant where the tests were performed is, depending on the format and characteristics of the brick manufactured, approximately 750 t/day. Such large test runs are necessary to assess with sufficient accuracy, considering the latency of the tunnel kiln, the effects of additions or substitutions on the firing process, emissions into air, and the energy consumption.

The use of ashes as an addition to the brick feed must take into account that in the clay free calcium ions are available for a pozzolanic reaction with some components of the biomass ash. This reaction is not immediate but rather slow and certainly accelerated by temperature. This reaction can cause problems in extrusion: stiffening associated with a loss of plasticity and hence denting of the extruded strand. It is hence necessary to hinder or delay this behavior either by adding the ashes, as done for polystyrene, for example, directly prior to extrusion or by hydrophobizing the ashes at least partially with a suitable agent. In brick making the two most suitable agents are waste glycerine from the production of biodiesel and waste antifreeze from the car industry. So far, no industrial tests with waste antifreeze compounds have been carried out. The trials were carried out with glycerine-treated ash.

The fruit and the wood combustion ashes were mixed in a 30:70 ratio and then subjected to hydrophobization with waste glycerine. The ashes were added to the raw material feed prior to its storage in the silo for souring. The raw material characteristics are indicated in Table 9.2. The mix was made up of 20% by volume of ashes and 80% by volume of clay. The use of ashes resulted in a noticeable reduction of the density and hence substantial improvement of the thermal performance of the finished product, as summarized in Table 9.3.

The reduction in unit weight is tied to a lower setting density. In the tests carried out, a reduction of about 8% of fuel was observed. A certain percentage can certainly be attributed to the energy content, body fuel, of the biomass ash.

Extrusion data were collected automatically and show that the power requirement of the extruder was lowered by 8%, whereas the density achieved is lower. The overall savings in production costs, considering that the ash is delivered free of charge to the brick plant, is estimated to be about 7.5%.

Table 9.2 Mixture of ash and brick fed used in this study

	Biomass ash	Brick feed
Density (kg/l)	0.3	1.73 raw density (not fired) In the plant a raw material mix with a fired density of 1.625 kg/l was used. This mix contains about 25% of nonquarried secondary raw materials and 75% of quarried materials
Humidity	1–2%	18.9%
Waste glycerine used as percentage of brick feed weight	0.25	
Waste glycerine used in kilograms per kilogram of biomass ash/brick feed mix	$1.73 \times 0.8 \times 0.25\% = 0.0035$ kg	
Waste glycerine used as percentage per kilogram of biomass ash	$0.0036/0.30 \times 0.2 = 5.7667\%$	

Table 9.3 Comparison of standard and ash-added bricks

	Standard brick	Ash-added brick
Unit weight	19.5	15.5
Material density of fired brick	1.625	1.475
R (m^2 K/W)	2.2	2.71
Thermal capacity (kJ/m^2 K)	>300	265
Acoustic Rw′R (dB)	>45	>43
Average/least compressive strength (N/mm^2)	14/10	9/7.5

Table 9.4 Results of the chemical analysis of bricks and ashes

		Unit	Fruit combustion ashes		Wood combustion ashes			Reference brick mix not fired	Brick mix with 20% combustion ashes
			Autumn	Early summer	Plant 1	Plant 2	Plant 3		
Aluminum oxide	Al_2O_3	% DS	3.7	3.9				6.5	
Sulfur trioxide	SO_3	% DS	3.8	2.8					0.70
Antimony oxide	Sb_2O_3	mg/kg DS	<1.2	2.9				<5.0	<1.2
Barium oxides	BaO	mg/kg DS	1.000	800				<5	650
Boron trioxide	B_2O_3	mg/kg DS	420	260				<500	82
Calcium oxide	CaO	% DS	6.9	9.2	29.4	28.9		<10	10.8
Iron oxide	Fe_2O_3	% DS	1.2	1.4	9.9	8.9		<10	6.3
Phosphorus pentoxide	P_2O_5	% DS	9.51	14.34				<0.5	1.21
Magnesium oxide	MgO	% DS	2.9	4.3	3.5	5.2		<5.0	2.4
Manganese oxide	MnO	% DS	0.04	0.05				<0.25	0.09
Lead oxide	PbO	mg/kg DS	22	41				<150	73
Potassium oxide	K_2O	% DS	14.9	15.5	8.9	1.2		<25	3.3
Silica	SiO_2	% DS	26.3	32.6	13.9	19.4		<65	39.5
Sodium oxide	Na_2O	% DS	0.1	1.4				<1.5	0.7
Titanium dioxide	TiO_2	mg/kg DS	1,400	1,800				<15,000	1,000
Zinc oxide	ZnO	mg/kg DS	230	600	812	788	705	<1,000	230
Zirconium oxide	ZrO_2	mg/kg DS	140	340				<450	290
Chlorides	Cl	mg/kg DS	<50	<50				<100	<50
Arsenic	As	mg/kg DS	2	<1		1.6			13
Cadmium	Cd	mg/kg DS	<1	1	11.4	5.3	3.9	<50	<1
Chromium	Cr	mg/kg DS	135	190	148.1	114.1	49.2	<200	140
Mercury	Hg	mg/kg DS	<1	<1	<1	<1	<1	<200	<1
Nickel	Ni	mg/kg DS	61	77	52	89.5	47.5	<2	
Copper	Cu	mg/kg DS	280	350	159	141	93	<100	
Selenium	Se	mg/kg DS	<2	<2				<500	

DS dry matter

Reactive swelling was not observed, at least at the percentages used in the tests. The bricks manufactured with the addition of the biomass ashes are of good quality. The analytical data for the raw ashes used and the brick manufactured are given in Table 9.4. No significant changes to the leaching values were found between a standard brick and a brick formed from the addition of biomass combustion ashes.

9.7 Conclusion

When hydrophobized biomass ashes were not used, the extrusion power requirements increased rapidly to a point where the tests had to be stopped because of excessive power requirement.

The results of the tests with hydrophobized ashes are as follows:

- Lower power requirement of the extruder at almost the same cutting frequency but at the same time a higher extrusion pressure
- Reduction of the water content of the brick by 2.5–3.0% (wt)
- Reduction of drying cracks
- Reduction of fired sulfate visible on the surface of the brick
- Substantial modifications to the firing curve

These results are positive. A problem, at least with the ashes used in the test, is that they are not delivered in a way that allows the brickyard to accept them, without any major and costly modifications to the plant. Most are delivered in big bags. This requires extra effort and expense on the part of the brickyard. If the ashes were delivered in trucks, an appropriate silo, similar to one used for coal or pet coke, could be installed. Another unresolved problem is the seasonality of the ashes. A storage bunker to overcome this problem usually cannot be accommodated in a brickyard. Once the producer of the ash has resolved these problems, a brickyard can certainly become an ideal recycler for this type of waste.

References

Anderson M, Skerratt G (2003) Variability study of incinerated sewage sludge ash in relation to future use in ceramic brick manufacture. Br Ceram Trans 102:109 ff

Berra M, de Casa G, Dell'Orso M, Galeotti L, Mangialardi T, Paolini AE, Piga L (2011) Reuse of woody biomass fly ash in cement-based materials: leaching tests. In: Insam H, Knapp BA (eds) Recycling of biomass ashes. Springer, Heidelberg, pp 133–146

Knapp BA, Insam H (2011) Recycling of biomass ashes – current technologies and future research needs. In: Insam H, Knapp BA (eds) Recycling of biomass ashes. Springer, Heidelberg, pp 1–16

Moedinger F (2003) Recycling and reuse of waste materials as raw materials within the framework of integrated product policy IPP. In: Dhir RK, Newlands MD, Dyer TD (eds) Recycling and reuse of waste materials. Thomas Telford, London, p 117 ff

Moedinger F (2004) Attenzione alle Sostenibilità – criteri e modalità per utilizzare rifiuti indisturiali nella produzione di laterizi. Riv Bioarchit 39–40:72–75

Moedinger F, D'Anna G (2002) Esperienze aziendali nel recupero dei rifiuti: condizioni contrattuali e procedure interne di controllo e gestione, L'Industria dei Laterizi XIIV:77–89

Omil B, Sánchez-Rodríguez F, Merino A (2011) Effects of ash applications on soil status, nutrition and growth of *Pinus radiata* D. Don plantations. In: Insam H, Knapp BA (eds) Recycling of biomass ashes. Springer, Heidelberg, pp 69–86

Ribbing CM, Bjurström HG (2011) The Swedish ash programme with focus on bio-ashes. Ashes are a resource in a sustainable society. In: Insam H, Knapp BA (eds) Recycling of biomass ashes. Springer, Heidelberg, pp 147–164

Chapter 10
Reuse of Woody Biomass Fly Ash in Cement-Based Materials: Leaching Tests

Mario Berra, Giancarlo De Casa, Marcello Dell'Orso, Luigi Galeotti, Teresa Mangialardi, Antonio Evangelista Paolini, and Luigi Piga

Abstract The feasibility of using woody biomass fly ash (WBFA) as a mineral admixture in cement-based materials was investigated. This fly ash was characterized for chemical composition and used to prepare a cement blend with 70 wt% Portland cement and 30 wt% WBFA. Cubic specimens were cast from a blended cement paste (water-to-binder ratio 0.50) and, after 28 days of curing at 20°C and 100% relative humidity, these specimens were tested for heavy metal leachability through the use of a sequential leaching protocol, at a constant pH of leachant (deionized water; pH 6.0). It was found that, except for the chloride content, the WBFA is able to meet the European chemical requirements established for reuse of coal fly ash in cement-based materials. Although the WBFA is characterized by a significant content of heavy metals of particular environmental concern (Cd, Cr, Cu, Ni, Pb, Zn), the results of the monolith leaching test have shown a good immobilization capacity of such metals by the cementitious matrix and, consequently, a good environmental quality of the blended cement investigated.

10.1 Introduction

In March 2007, the European Commission undertook an approach to climate and energy policy in order to fight climate change and increase EU energy security while strengthening its competitiveness. The European Commission committed itself to transform Europe into a highly energy efficient, low-carbon economy.

M. Berra
ERSE S.p.A., Milan, Italy

G. De Casa, L. Galeotti, T. Mangialardi, A.E. Paolini, and L. Piga (✉)
Faculty of Engineering, Sapienza University of Roma, Rome, Italy
e-mail: luigi.piga@uniroma1.it

M. Dell'Orso
Chemical Laboratory of DGERM-Economic Development Ministry, Rome, Italy

H. Insam and B.A. Knapp (eds.), *Recycling of Biomass Ashes*,
DOI 10.1007/978-3-642-19354-5_10, © Springer-Verlag Berlin Heidelberg 2011

To achieve this goal, the European Commission aimed to carry out, by 2020, what is known as the 20:20:20 project, namely:

- A reduction in EU greenhouse gas emissions of at least 20% below 1990 levels
- Twenty percent of EU energy to be produced from renewable resources
- A 20% reduction in primary energy use to be achieved by improving energy efficiency

In this context, the use of biomasses in place of traditional fuels represents a suitable way of reducing greenhouse gas emissions. In fact, the biomasses may be regarded as clean and renewable energy resources with no net CO_2 production, since the amount of CO_2 produced from biomass combustion is approximately equivalent to that taken up from the environment during biomass growth.

The most important biomasses are the residues from woodworking or forest activities, the wastes from farms and agrobusiness, the organic fraction of municipal solid wastes, and the plants deliberately grown for energy production purposes. In Italy, the most utilized biomasses for burning in power plants are chipped wood, and, to a minor extent, rice-husk and olive residues (GSE 2009).

Although the use of biomass in Italy is less than the European average, the high potential of burnable biomass along with the fast increase in the number of biomass-based thermal plants calls for a solution to the disposal problems associated with ash production. Both the quality and the quantity of ash depend on the type of biomass used as a fuel. The amount of ash produced per unit weight of original biomass can vary from about 2% (w/w) (chipped wood) to 15% (w/w) (rice husk) (Lokare et al. 2006).

Irrespective of the type of biomass used, two kinds of ashes are produced: fly ash and bottom ash. Fly ash is generally trapped by electrostatic precipitators or sleeve filters located downstream of the combustion process, before the gas and the very fine particles are released to the environment. Bottom ash is collected in the bottom of the boiler. The relative amount of fly ash and bottom ash depends on the type of boiler. Powder boilers produce more bottom ash than fly ash; fluidized bed boilers produce more fly ash than bottom ash. Grate boilers produce about the same quantity of both ashes.

According to the European waste catalog and hazardous residues list (Commission of the European Communities 2000), both fly ash and bottom ash originating from combustion of untreated wood are classified as nonhazardous wastes and are listed with codes 10.01.03 and 10.01.01, respectively. The former code also includes fly ash from peat; the latter also includes slag and boiler dust.

Woody biomass ash, being a waste, has to be disposed of in authorized landfills. Alternatively, this waste may be reused as a fertilizer or for building purposes, provided it passes the tests prescribed by the environmental laws. Bottom ash may be used directly as a building material to replace granular material in geotechnical works, such as road foundations. Fly ash may be reused as a filler in cementitious mixes. However, the high content of alkalies and chlorides could prevent the reuse of fly ash in cementitious mixes.

From the environmental point of view, reuse of biomass fly ash in concrete would be very profitable as partial replacement of Portland cement. This may (1) solve the problem of fly ash disposal, (2) reduce the CO_2 emissions involved in the industrial production of cement from traditional raw materials, namely, limestone and clay (0.83 t CO_2 is emitted for each ton of Portland cement produced), and (3) preserve the natural resources involved in cement production, with further beneficial effects on the environment.

A recent study (Rajamma et al. 2009) has shown that the replacement of Portland cement with woody biomass fly ash (WBFA) up to 20% (w/w) of cement does not negatively affect the development of the mechanical properties of cementitious mixes. The practical inference of such fly ash reuse would be a 20% reduction of CO_2 emission related to cement production, and this would be an innovation in line with what is expected by the European Commission.

However, the reuse of biomass fly ashes in cement-based materials is strongly related to their chemical and environmental characteristics. Generally, fly ashes originating from traditional and innovative fuels may contain significant amounts of heavy metals that pose severe limitations for their disposal in landfills or for their reuse in agricultural/industrial applications. In particular, cadmium appears to be the most problematic heavy metal in biomass fly ashes, and chromium, mainly in the Cr(VI) state, may be problematic in many ash stabilization processes because of its mobility at high pH values (Lima et al. 2008). Presently, little it is known about the environmental compatibility of blended cements made with Portland cement and biomass fly ash.

In this study, the leaching behavior of a mixture of Portland cement and WBFA was investigated in view of the possible reuse of this kind of fly ash as a mineral admixture in the formulation of blended cements. The eco-compatibility of such a mixture was assessed through the use of a monolith leaching test on hardened cement pastes under constant pH conditions (pH 6.0).

10.2 Materials and Methods

The WBFA used in this study came from the electrostatic precipitators of an Italian chipped-wood burning plant. The sample of WBFA "as received" contained less than 1% (w/w) moisture and about 30% (w/w) of particles with sizes above 150 μm. This fly ash was first sieved with a 150-μm sieve, and the retained portion was ground to fineness below 150 μm and then mixed with the remaining ash portion. The characteristic diameters of the resulting powder were as follows: $D_{10} = 1.34$ μm, $D_{50} = 11.99$ μm, and $D_{90} = 38.86$ μm. This powder was used to prepare the blended cement pastes to be subjected to monolith leaching tests.

The X-ray diffraction pattern (not reported here) showed that the WBFA mainly consisted of a vitreous matrix, with the presence of crystalline phases such as lime, calcite, and quartz.

Table 10.1 Elemental composition of woody biomass fly ash (WBFA)

Element	Dry solid (% w/w)	Element	Dry solid (mg/kg)
Ca	15.09	Zn	2,274
Si	13.01	Pb	177
Al	4.49	Cu	175
Fe	3.98	Cr	101
K	3.02	Ni	41
Mg	2.1	Cd	9
Na	0.98		
P	0.73		
S[a]	1.16		
Cl	1.07		
LOI at 950°C[b]	4.50		

[a]Acid-soluble sulfate expressed as sulfur
[b]loss on ignition

Table 10.1 gives the elemental analysis of WBFA, as determined by X-ray fluorescence for major elements and by atomic absorption spectrophotometry (AAS) for trace metals. Sulfates and chlorides were determined by ionic chromatography (high-performance liquid chromatography, HPLC). The contents of unburned carbon and inorganic carbon of fly ash, as determined by simultaneous thermogravimetry/differential scanning calorimetry (static air; heating rate of 10°C/min over a temperature range from 25 to 1,000°C), were 0.5% (w/w) (weight loss in the range of 300–500°C) and 1.09% (w/w) (weight loss in the range of 630–750°C), respectively.

A sample of WBFA was subjected to elution with deionized water to evaluate the contents of water-soluble species, such as alkalies, chlorides, and sulfates, as well as the pH and the percentage of ash dissolution.

The elution test was carried out according to the experimental procedure specified in the European leaching test UNI-EN 12457-2. In particular, an aqueous suspension of fly ash (liquid-to-solid ratio of 10 L/kg) was put in a closed vessel and left for 24 h under agitation at 20°C. Afterwards, the suspension was allowed to settle for 15 min and was then vacuum-filtered through a 0.45-μm membrane filter, with no rinsing of the filter. Finally, the eluate was analyzed for the concentrations of chloride and sulfate ions (by HPLC) and sodium and potassium ions (by AAS), as well as for the electric conductivity (by a conductimeter) and pH (by a pH meter). The weight loss of fly ash was also determined after drying of the solid residue to a constant weight at 80°C.

10.2.1 Preparation of Hardened Cement Paste Specimens for Monolith Leaching Test

A sample of WBFA was dry-mixed with Portland cement (CEM I 42.5R) with an ash-to-cement ratio of 30:70 by mass, and the resulting blended cement (binder) was used for environmental compatibility studies. Table 10.2 gives the chemical

Table 10.2 Chemical and mineralogical compositions of Portland cement

Component	Percentage (w/w)	Bogue potential constituent[a]	Percentage (w/w)
CaO	63.0	C_3S	54.6
Free CaO	0.70	C_2S	16.3
SiO_2	20.0	C_3A	9.23
Al_2O_3	5.04	C_4AF	7.42
Fe_2O_3	2.44		
MgO	1.35		
Na_2O	0.30		
K_2O	1.12		
P_2O_5	0.13		
SO_3	3.36		
Cl	0.01		
LOI at 950°C	1.74		

[a]Cement chemistry notation is used: C is CaO; S is SiO_2; A is Al_2O_3; F is Fe_2O_3.

and mineralogical compositions of the Portland cement, the composition of the latter being calculated by the Bogue method.

Cement pastes were prepared manually by using the blended cement and deionized water as mixing water, at a water-to-binder weight ratio (w/b) of 0.50.

Cubic specimens, with sides of 40 mm, were cast from the cement paste and, after 24 h of curing within the molds, the specimens were demolded and cured for 28 days in a controlled temperature and humidity environment (20°C and relative humidity above 95%). Afterwards, the cubic specimens were subjected to the monolith leaching test.

10.2.2 Monolith Leaching Test

The monolith leaching test used in this study was a modified version of the conventional NEN 7345 leaching test (Dutch Standardization Institute 1995).

The conventional NEN 7345 test consists of immersing the test specimen in a closed polypropylene reactor containing deionized water as a leachant (initially acidified to pH 4.0 by nitric acid addition). A liquid-to-solid volume ratio (L/S) of 5.0 was used. During the test, there was no agitation of the leaching solution and the pH was free to change according to the acid or basic characteristics of the compounds released from the test specimen. The leachate was periodically replaced with an equal volume of leachant, after cumulative leach times of 0.25, 1, 2.25, 4, 9, 16, 36, and 64 days. The eight leachates were filtered through a 0.45-μm membrane filter and, after pH and electric conductivity measurements had been conducted, they were acidified to pH 2.0 with 5 M nitric acid. Finally, each leachate was analyzed for the concentrations of selected heavy metals (in this study, Cd, Cu, Cr, Ni, Pb, and Zn) by using an atomic absorption spectrophotometer equipped with a graphite furnace.

The modified leaching test used in this study differed from the conventional NEN 7345 test only for the continuous and moderate agitation of the leaching solution within the reactor (realized by a magnetic stirrer) and, especially, for the pH evolution throughout the test. In particular, the pH was kept constant to a predetermined value by means of an automatic titrator. The pH was monitored constantly, and the titrator compensated for alterations of pH by small additions of 0.5 M HNO_3 from a burette. A pH of 6.0, which is characteristic for most natural waters, was selected for the modified NEN 7345 test. The addition of nitric acid was motivated by the necessity of compensating for the pH rise due to the release of alkaline compounds (alkalies and calcium hydroxide) from the cementitious test specimen, without altering the leachability of heavy metals through complexation reactions. Indeed, the nitrate ion is known as a poor complexing agent toward most heavy metals. Figure 10.1 illustrates the apparatus used for the monolith leaching test.

Fig. 10.1 Apparatus used for the monolith leaching test

Three replicate leaching tests were performed and the test results were averaged. In each test, one cubic specimen (64 cm^3 in volume) was used and was moistened with 320 cm^3 of deionized water (L/S = 5.0; eight leachant renewals).

10.3 Results and Discussion

10.3.1 Chemical Characterization of WBFA

As shown in Table 10.1, the WBFA was characterized by a significant presence of heavy metals of particular environmental concern, such as cadmium, chromium, copper, lead, and zinc. Zinc was the predominant heavy metal (2,274 mg/kg), whereas cadmium was the heavy metal with the lowest concentration (9 mg/kg).

With regard to the reuse of WBFA in cement-based materials, it must also be considered that several other chemical species, such as chlorides, sulfates, alkalies, and magnesium oxide contained in this waste (Table 10.1), may exert adverse effects on cement hydration and concrete durability.

It is undoubted that the presence of a significant amount of water-soluble compounds, such as chlorides and alkalies, could promote the formation of a high porosity within the hardened cementitious mixes, thus penalizing mechanical strength development and durability. High contents of water-soluble chlorides can also be deleterious for steel-reinforced concrete, since they will promote the corrosion of iron reinforcing bars. High contents of available alkalies (i.e., the amount of alkalies released into the pore liquid of cementitious matrices) are responsible for the development of deleterious expansion associated with alkali–silica reaction in concretes made with aggregates containing some alkali-reactive forms of silica or silicate, or both. Deleterious expansive phenomena in concrete can also arise from very slow dissolution of significant amounts of sulfates or magnesium oxide, with subsequent precipitation of very expansive phases, such as ettringite ($CaSO_4 \cdot 32H_2O$) and brucite [$Mg(OH)_2$].

In the USA, the current restrictions concern the use of fly ashes originating from combustion of fossil fuels (bituminous and subbituminous coal, peat, and lignite), and these restrictions are specified in the ASTM C 618 method. In Europe, the current restrictions concern both the fly ashes derived from fossil fuels and those resulting from the combustion of biomass and fossil fuel blends (cofiring), with a biomass content not higher than 20 wt%. These restrictions are specified in the EN 450-1 method.

For the above reasons there exists no standard method dealing with the chemical requirements for reuse of fly ash from pure biomass combustion in cementitious mixes. However, it is reasonable that the quality of fly ash from pure biomass burning should follow the same regulation as fly ash from fossil fuel combustion.

Table 10.3 compares the specific chemical characteristics of the WBFA (also reported in Table 10.1 in terms of elemental composition) with the corresponding

Table 10.3 Comparison between the chemical characteristics of WBFA and the chemical requirements for reuse of fly ash in concrete

	Chemical characteristic (wt%)		Chemical requirement	
	WBFA	Washed WBFA	EN 450-1	ASTM C618
LOI	4.5	8.1	9.0[a]	6.0
Chloride	1.07	0.20	0.10	–
Sulfate (as SO_3)	2.9	0.75	3.0	5.0
Alkalies (as Na_2Oeq[b])	3.72	1.43	5.0[c]	1.5[d]
Free calcium oxide	2.4	–	2.5	–
Magnesium oxide	3.5	3.9	4.0	–
Water-soluble phosphate (as P_2O_5)	0.003	–	0.01	–

[a]Category C
[b]$\%Na_2Oeq = \%Na_2O + \%K_2O \cdot 0.66$
[c]Acid-soluble alkalies (EN 196-2)
[d]Available alkalies (ASTM C311)

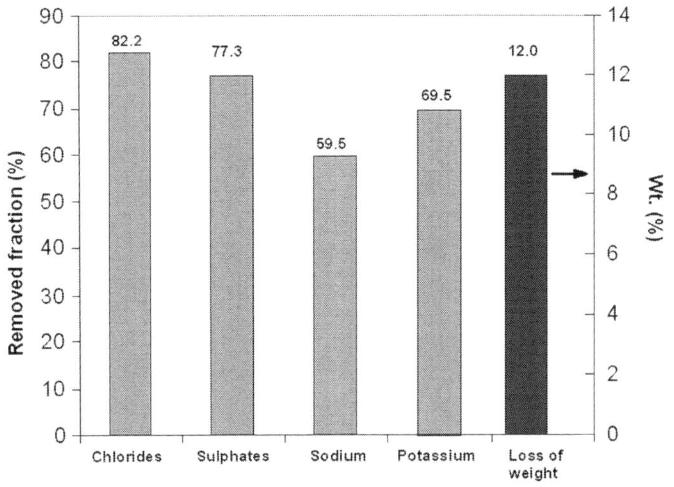

Fig. 10.2 Results of the elution test on the woody biomass fly ash sample

limits (chemical requirements) specified by ASTM C 618 and EN 450-1. In this table, the chemical characteristics of the solid residue resulting from the elution of WBFA with deionized water (elution test) are also reported.

The chemical characteristics of this solid residue, also referred to as washed WBFA, were evaluated on the basis of the results of the elution test reported in Fig. 10.2, in terms of percentage removals of alkalies, chlorides and sulfates, and percentage weight loss of WBFA.

As shown in Table 10.3, with exception made for the loss on ignition, the American specifications appear to be less severe than the European specifications. In particular, ASTM C 618 does not restrict chloride, free calcium oxide, magnesium oxide, and soluble phosphate contents. Moreover, the sulfate limit (5%) is more than the maximum allowable (3%) by European regulation. As far as the alkali content is concerned, the American limit [1.5% (w/w) as available alkalies] appears to be comparable to the European limit [5.0% (w/w) as acid-soluble alkalies] if it is considered that, depending on the type of coal fly ash, the content of available alkalies, as determined by the ASTM C311 test method, may vary from 20 to 50% of total alkalies (acid-soluble alkalies), as determined by the EN 196-2 test method (Berra et al. 1992).

According to the chemical requirements prescribed by EN 450-1, which are mostly more severe than those required by ASTM C 618, and remembering that, to date, these requirments do not apply to pure biomass fly ashes, the WBFA was found to meet all chemical requirements, except for the chloride content (1.07% against the limit of 0.10%) (Table 10.3).

In spite of the high release of water-soluble chlorides (82.2% removal) accompanied by a relatively low percentage weight loss of fly ash (12%) (Fig. 10.2), even the washed WBFA failed to meet the chemical requirement for chlorides, but the chloride content (0.20%) was only slightly higher than the limit of 0.10% (w/w) (Table 10.3).

As compared with WBFA, washed WBFA was also characterized by much lower contents of alkalies [1.43% (w/w) Na_2Oeq against 3.7%] and sulfates [0.75% (w/w) SO_3 against 2.9%]. Conversely, washed WBFA was more rich in MgO [3.9% (w/w) against 3.5% for WBFA], but the MgO content remained below the limit of 4.0% (Table 10.3). The washed WBFA could also be richer in heavy metals than WBFA, in consideration of the expected low release of these substances into aqueous solutions.

As far as the use of washed WBFA in cement-based materials is concerned, it is likely that a washing treatment of fly ash with a liquid-to-solid ratio above 10 L/kg could reduce the chloride content below the 0.1% limit (Table 10.3). In that case, the wastewater resulting from the washing of fly ash could contain chloride and sulfate concentrations below the limits established for wastewater disposal. However, this wastewater should be treated for pH correction and, probably, for heavy metal removal. In this regard, the pH of WBFA, defined as the pH of the aqueous suspension of fly ash with an L/S ratio of 10 L/kg, was 12.9. This high pH, which is compatible with the use of WBFA in cement-based materials, was attributable to the release of alkalies and calcium oxide from fly ash.

The electrical conductivity of the eluate (1.53 S/m) evidenced significant release of electrolytic compounds from WBFA.

10.3.2 Leaching Behavior of Blended Cement Pastes

Table 10.4 gives the results of the monolith leaching test on cubic specimens of cement pastes (water-to-binder weight ratio 0.50) made with blended cement [70% (w/w) Portland cement and 30% (w/w) WBFA]. In this table, the concentrations of selected heavy metals (Cd, Cr, Cu, Ni, Pb, and Zn) in each of eight leachates are reported as the average values of three replicate leaching tests.

Table 10.4 Results of the monolith leaching tests on cubic specimens of blended cement pastes

Renewal number	Heavy metal concentration (μg/L)					
	Cu	Cd	Ni	Pb	Cr	Zn
1	20.9	0.7	5.3	46.6	5.4	105
2	10.8	0.4	3.3	8.1	0.5	100
3	16.2	0.4	2.1	1.2	1.2	46
4	10.2	0.5	0.3	7.0	0.3	72
5	13.0	0.5	4.4	3.6	3.2	93
6	11.4	0.5	8.5	1.8	2.5	34
7	16.4	0.5	1.5	14.9	8.8	135
8	13.0	0.5	2.0	10.0	4.0	105

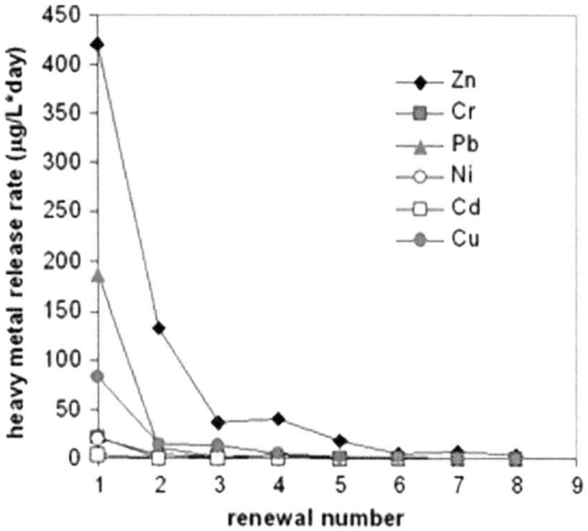

Fig. 10.3 Average release rates of heavy metals within the leaching periods

With respect to the other heavy metals investigated, the higher concentrations of copper, lead, and zinc measured for most leachates were directly related to their higher contents in the original fly ash (Table 10.1).

Using the data in Table 10.4, we calculated the leaching rate of each of the selected heavy metals as an average within each leaching period, and these rates are reported in Fig. 10.3 for each of the eight leachant renewals.

For copper, lead, and zinc, the leaching rates dramatically reduced after the first leachant renewal (the first two renewals for Zn), thus revealing the existence of two different mechanisms governing the leaching process of such heavy metals. At early leaching times (first two renewals), the controlling mechanism appeared to be the release of heavy metal from the outer surface of the monolith specimen by dissolution into the leaching solution or by wash-off, or both. At longer leaching times, the release was probably controlled by diffusion, and the heavy metal ions had to migrate within the pore liquid of the cementitious matrix of the test specimen prior to reaching the liquid bulk. As a result, this leaching phase was characterized by a much lower rate as compared with the initial leaching phase. In the case of cadmium, chromium, and nickel leaching, no dissolution/wash-off phenomenon was detected during the early release phase.

As shown in Fig. 10.3, after the first or the second leachant renewal, the release rate of each heavy metal did not significantly vary with increasing leaching time. Thus, the high concentrations of heavy metals measured for the seventh and eighth leachates (Table 10.4) were attributable to the much higher contact times between the specimen and the leachant (20 and 28 days for the seventh and eighth renewals, respectively).

With use of the results in Table 10.4, the cumulative mass of each heavy metal released per unit exposed surface area of specimen, M_t (mg/m^2), was also calculated and is plotted in Fig. 10.4 as a function of the square root of the cumulative leach time, t (h$^{1/2}$).

For the leaching of cadmium, chromium, and nickel, there existed straight line relationships between M_t and $t^{1/2}$, with no intercept on the coordinate axis. This is typical of leaching processes controlled by the diffusion mechanism. Conversely, for copper, lead, and zinc leaching, linear relationships with positive intercepts on the ordinate axis were obtained. This is typical of leaching processes controlled initially by dissolution or wash-off phenomena, or both.

To predict the long-term release of copper, lead, and zinc from monolithic specimens, the leaching data in Table 10.4, relative to these metals, were considered over the leach time interval for which diffusion was the release-controlling mechanism. In other words, the first two leachant renewals were considered as preconditioning steps for the subsequent leaching test. In this way, straight line relationships between M_t and $t^{1/2}$ were obtained for the release of copper, lead, and zinc, as shown in Fig. 10.5.

With use of the linear regression equations resulting from the data in Figs. 10.4 and 10.5, the releases of heavy metals after 100 years of leaching were estimated

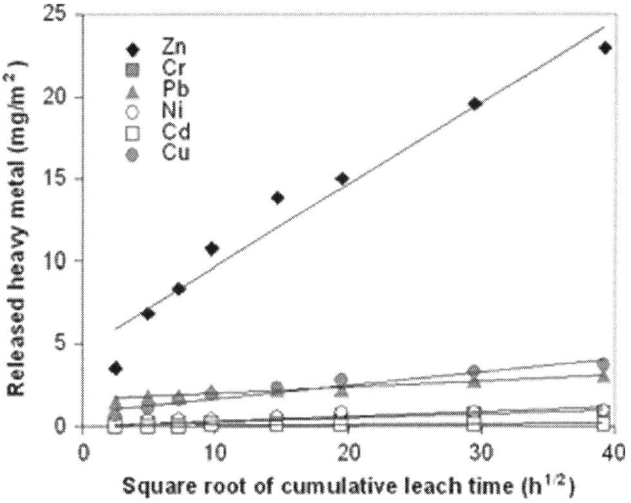

Fig. 10.4 Cumulative release of heavy metals as a function of the square root of leaching time

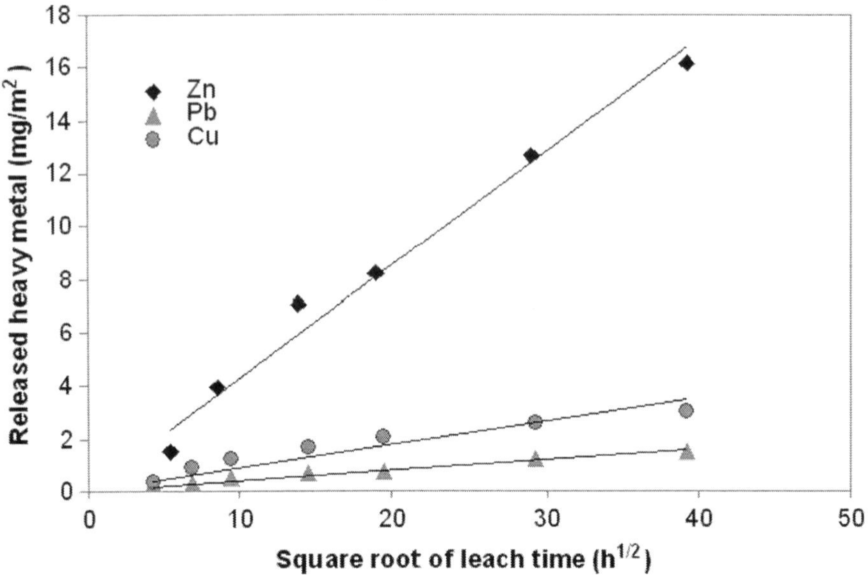

Fig. 10.5 Cumulative release of Cu, Pb, and Zn as a function of the square root of leach time (diffusion-controlled leaching data)

and compared with the standard limits (Category I applications) as specified in the Dutch Building Materials Decree (1995). These specifications are commonly taken as a reference for evaluating the environmental quality of cement-based materials incorporating hazardous wastes. Figure 10.6 compares the estimated releases of the selected heavy metals with the Dutch standard limits.

As can be noted, all the releases were well below the corresponding regulatory limits and this proved the good immobilization capacity of heavy metals by the cementitious matrix investigated and, consequently, the good environmental quality of the blended cement formulated with 30% (w/w) WBFA.

10.4 Conclusions

The WBFA is characterized by a significant content of heavy metals of particular environmental concern, such as cadmium, chromium, copper, nickel, lead, and zinc, and by a remarkable amount of water-soluble compounds, such as alkalies, chlorides, and sulfates.

According to the European chemical requirements established for reuse of coal fly ash as a mineral admixture in cement-based materials, the biomass fly ash studied appears to be suitable for the formulation of blended cements, provided that its chloride content be preliminarily reduced.

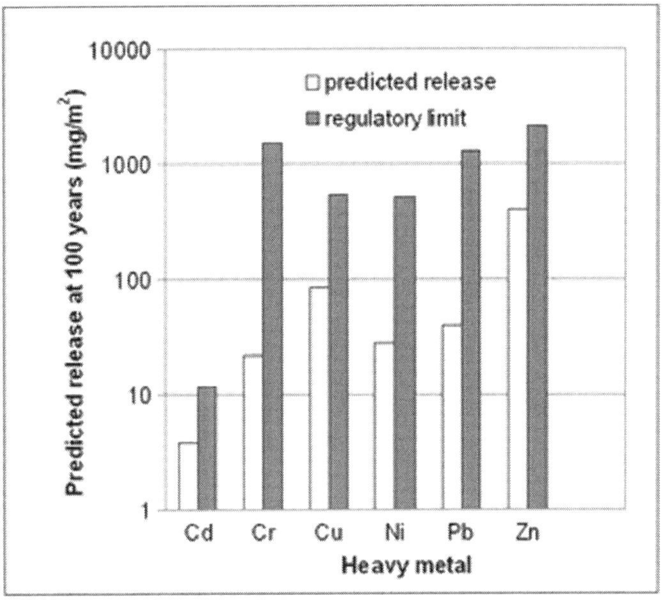

Fig. 10.6 Prediction of long-term release of heavy metals

As indicated by the results of the water elution test on WBFA, a single-stage washing treatment of this ash with deionized water might be sufficient to reduce the chloride content to acceptable levels.

As evidenced by the results of the monolith leaching test on hardened pastes of blended cement [70% (w/w) Portland cement–30% (w/w) WBFA], in spite of the high content of water-soluble compounds of WBFA and the acid pH conditions of the leachant throughout the test (pH 6.0), very low releases of heavy metals were always obtained, thus revealing a high metal immobilization capacity by the cementitious matrix and, consequently, a good environmental quality of the blended cement investigated.

For some heavy metals such as copper, lead, and zinc, the release from a monolithic specimen appears to be governed by two different leaching mechanisms: dissolution/wash-off at earlier leach times and diffusion at longer leaching times. Conversely, in the case of cadmium, chromium, and nickel leaching, no dissolution/wash-off phenomenon was detected.

References

Berra M, Mangialardi T, Paolini AE, Turriziani R (1992) Effect of fly ash on alkali-silica reaction. In: Poole AB (ed) Proceedings of the 9th international conference on alkali-silica reaction in concrete, vol I. Concrete Society Publication, London, pp 61–70

Building Materials Decree (1995), bouwstoffenbesluit, nr. 12689/165, The Netherlands Ministry of Housing, Spatial Planning and the Environment VROM

Commission of the European Communities (2000) Establishing a list of wastes pursuant to Article 1(a) of Council Directive 75/442/EEC on waste and Council Decision 94/904/EC establishing a list of hazardous waste pursuant to Article 1(4) of Council Directive 91/689/EEC on hazardous waste. Official Journal of the European Communities L 226/3

Dutch Standardization Institute (1995) NEN 7345 Leaching characteristics of soil and stony building and waste materials – leaching tests – Determination of the leaching of inorganic components from building and monolithic materials with the diffusion test

GSE, Gestore Servizi Elettrici (2009) "Le biomasse e i rifiuti – Dati statistici al 31/12/2008" Ufficio statistiche. http://www.gse.it/attivita/statistiche/Documents/Biomasse%20e%20Rifiuti%202009.pdf

Lima AT, Ottosen LM, Pedersen AJ, Ribeiro AB (2008) Characterization of fly ash from bio and municipal waste. Biomass Bioenergy 32:277–282

Lokare SS, Dunaway JD, Moulton D, Rogers D, Tree DR, Baxter LL (2006) Investigation of ash deposition rates for a suite of biomass fuels and fuel blends. Energy Fuel 20:1008–1014

Rajamma R, Ball RJ, Tarelho LAC, Allen GC, Labrincha JA, Ferreira VM (2009) Characterisation and use of biomass fly ash in cement-based materials. J Hazard Mater 172:1049–1060

Chapter 11
The Swedish Ash Programme with Focus on Bioashes: Ashes Are a Resource in a Sustainable Society

Claes M. Ribbing and Henrik G. Bjurström

Abstract The Swedish Ash Programme is an applied R&D programme aimed at demonstrating uses for combustion residues (ash) and providing an improved understanding of these residues for the purpose of resolving regulatory questions. Fuels are biomass, wastes, peat – any solid fuel but coal. The progress in the Ash Programme since its inception in 2002 is reviewed. The hierarchy for biomass ash is recycling to forest soils as compensation for the removal of mineral nutrients first, and use in civil works second. Assessment of the environmental impact in view of permitting procedures for civil works and ecotoxicity are particularly addressed.

11.1 Introduction

The Swedish Ash Programme is an applied R&D programme aimed at demonstrating uses for combustion residues (ash) and providing an improved understanding of these residues for the purpose of resolving regulatory questions. It is a collaborative undertaking implemented since 2002 by Värmeforsk, the Swedish Thermal Engineering Research Institute, and co-financed by the ash producers, i.e. the combustion plants, and the government, principally through the Swedish Energy Agency. The Swedish Environmental Protection Agency and the Swedish Road Administration also contribute financially.

The vision moving the Ash Programme is:

> "Combustion residues are resources in a sustainable society"

Since its inception in 2002, the Ash Programme has supported more than 100 applied R&D projects, most of them co-financed by other organisations. Including

C.M. Ribbing (✉)
Svenska Energiaskor AB, Torsgatan 12, 111 23 Stockholm, Sweden
e-mail: claes.ribbing@energiaskor.se

H.G. Bjurström
ÅF-Engineering AB, 169 99 Stockholm, Sweden
e-mail: henrik.bjurstrom@afconsult.com

H. Insam and B.A. Knapp (eds.), *Recycling of Biomass Ashes*,
DOI 10.1007/978-3-642-19354-5_11,

currently ongoing projects to be concluded by the end of 2011, they represent an investment of approximately €9 million. All of these are short actions aimed directly at a specific question, demonstrating on a large scale the utilisation of combustion residues or monitoring the environmental impact of a large-scale application. The programme does not support traditional university research over a period of several years.

The results achieved in the Ash Programme between 2002 and 2008 were reviewed in a contribution to the 2009 International Waste Management and Landfill Symposium (Bjurström et al. 2009). They are also described in a synthesis in English available from Värmeforsk's Web site (Bjurström and Herbert 2009). These results will be summarised briefly in this review, as a background to the themes focused on here, ash from solid biofuels and the regulatory process, which will be developed in more detail.

The areas of use targeted by the Ash Programme are (1) as a geotechnical material, e.g. in roads or other civil works, (2) in landfill construction and closure and (3) as mineral nutrients in wood ash recycled to forest soils. Issues common to all these areas are the chemistry of ash and environmental aspects.

The results obtained and the conclusions presented within the projects are those of the scientists. Environmental authorities do not automatically agree with the conclusions. To be more specific, bones of contention are the official environmental target "A non-toxic environment" and whether considering wastes as a resource is politically correct.

11.2 Combustion Residues: A Background

The fuels in Sweden are solid biofuels as well as different combustible wastes, including municipal solid wastes (MSW). The share of fossil fuels, coal or oil, is very small. The ash content in the fuels differs considerably. Clean heartwood as in wood pellets has an ash content of 0.2–0.5%. Wood chips obtained from forest residues have an ash content of 2–4% as the proportion of bark increases. Waste wood, such as construction and demolition debris, usually has higher ash contents. For peat, a figure of 5% is generally used, although the content can be much higher. Waste materials have comparatively high ash contents: 25% in MSW is quite common and various sludges in the pulp and paper industry have ash contents ranging between 10 and 50%.

11.2.1 Production of Combustion Residues

In 2006, 1.3 million tonnes of combustion residues was produced (estimate for 2008: 1.5 million tonnes) and almost 80% was utilised. A summary of the quantities and types of combustion residues is provided in Table 11.1. The presentation is split into more categories than is usual in surveys, because mixtures of fuels, some of

Table 11.1 Production of combustion residues in Sweden in 2006, tonnes of dry substances (DS) per year (survey performed by Svenska Energiaskor)

Type of furnace	Fuel	Bottom ash	Fly ash and APC residues	Combined bottom and fly ash
Grate furnaces and others	Municipal solid waste, industrial waste	445,000	93,000	–
	Solid biofuels (wood chips, logging residues)	13,000	10,000	77,000
	Solid biofuels and sludge from the pulp and paper industry	12,000	11,000	10,000
	Mixed fuels	35,000	25,000	–
	Wood chips and peat	2,500	1,000	2,000
	Coal	3,000	2,000	–
Pulverised fuel furnaces	Coal and peat	7,000	24,000	–
	Peat, wood, etc.	18,000	31,000	–
Fluidised bed furnaces (CFB, BFB)	Municipal solid waste, industrial waste	45,000	57,000	–
	Solid biofuels	15,000	14,000	5,000
	Mixed fuels	31,000	88,000	14,000
	Peat and wood chips	11,000	27,000	
	Pulp and paper industry	21,000	78,000	10,000
	Coal (PFBC)	5,000	44,000	
Subtotals	–	660,000	500,000	120,000
Grand total	1,280,000	–	–	–

APC air pollution control; *CFB* circulating fluidised bed; *BFB* bubling fluidised bed; *PFBC* pressurised fluidised bed combustion

them particular to a type of industry and to type of furnace, are important for the properties of the residues.

As one may see, the problem for a provider of materials is that the sources are numerous and small. For example, the 600,000 t/year of MSW incineration (MSWI) residues is produced by more than 25 plants. Many district heating plants produce less than 2,000 t of mostly wood-based ash per year. The smallest ones, as well as small sawmills, do not produce more than 1 t/year. The really small capacity furnaces are not included in these figures, e.g. pellet furnaces in individual homes, or farm units firing agricultural residues.

If one sums up all categories of solid biofuels and mixtures, the total quantity of ash from solid biofuels is of the order of 370,000 t/year.

All types of furnaces are used, grate furnaces, pulverised fuel (PF) furnaces and fluidised bed furnaces, the latter being perhaps more common in Sweden than in the rest of Europe. The capacities range from a few hundred kilowatts to a couple of hundred megawatts on a fuel basis. Small furnaces up to 10 MW fuel are usually grate furnaces, and fluidised bed furnaces are preferred from 20 MW fuel and upwards. PF furnaces are not so common in Sweden. All these types of furnaces have their particularities, which affect the properties of the residues; see Sect. 11.2.4.

11.2.2 The Definition of Biomass

In the EN 14588 standard, biomass is defined as "material of biological origin excluding material embedded in geological formations and transformed to fossil". When discussing solid biofuels, one generally assumes that the fuel has to have 100% biological origin and that it is a virgin material, not a waste. A few points should be made here.

Peat is normally regarded as something between a renewable and a fossil solid biofuel. From our point of view, peat is a biomass because:

It agrees with the definition in the standard.
The annual formation of peat in Sweden and Finland is much larger than the yearly harvest.
Vast peat land areas have already been drained in Sweden and in Finland, and these areas now leak climate change gases; utilising the energy that the oxidation releases is much better than not using it at all.
Harvesting old peatlands increases the rate of growth of new peat.

This being said, peat is a minor fuel: woody biomass represents 48% of the fuel supplied to district heating systems in Sweden, on an energy basis. Peat represents only 5%, but it is often used for its comparatively high sulphur content as an auxiliary fuel to abate corrosion.

Unless the biomass is harvested solely for energy and used as is, most solid biofuels are actually wastes. For example, logging residues are considered as virgin biomass although technically they are residues from the conventional exploitation of forests. Analogously, residues from sawmills, board production, and pulp and paper mills combusted at the mills for the energy needs of the mills are somewhere between virgin biomass and wastes. Waste biomass is still biomass according to EN14588. For example, often 85% of the energy in MSW is biomass, and industrial biomass waste contains normally somewhat less than that.

Incineration is destruction of waste without utilisation of the energy produced. This does not occur in Sweden, where all waste-burning plants provide heat to district heating systems and most of them are cogeneration plants. The Swedish High Environmental Court wanted to regard all combustion of industrial or municipal waste as incineration. It has, however, been overruled by the Advocate General of the EU: if the main purpose is to generate energy, then this is co-combustion, even if all fuel fractions are wastes.

To keep within the purpose of this review of the Ash Programme, we will not consider residues from the combustion of MSW. The review deals with other non-fossil fuels, i.e. those originating from wood or peat. The properties of these residues are discussed in Sect. 11.2.4.

11.2.3 Utilisations

The large quantities of combustion residues may be swallowed only by a mass market, that of materials for civil works. The largest single use in 2006 was as a

Table 11.2 The uses of combustion residues in Sweden in 2006 (survey performed by Svenska Energiaskor)

Area of use	Quantities (tonnes DS per year)
Landfill construction and closure	650,000
Civil works outside of landfills	200,000
Backfilling cavities (e.g. mines and quarries)	50,000
Spreading to forest soil and arable land	35,000
Other uses and unknown uses	175,000
Total quantity used	1,000,000
Total quantity produced	1,300,000

construction material in the closure of landfills (as well as for capping), with approximately 650,000 t; see Table 11.2. The financial incentive for this use is the possibility to waive the tax on waste sent to landfill: the materials are used and replace virgin materials. However, many currently active landfills will be closed within the next 10–15 years. It should be noted that backfilling cavities in Table 11.2 concerns mostly air pollution control (APC) residues from combustion of MSW.

Spreading to forest soils is a small area of use, with approximately 35,000 t/year, but is of vital importance for the sustainability of the production of solid biofuels from biomass harvested from forests. The relevant biomass fraction is the logging residues, and harvesting it on top of the extraction of timber and pulping wood in conventional forestry not only removes the mineral nutrients in the residues that if left in the forest would have been available to the next generation, but also exacerbates the natural acidification of forest soils by conventional forestry.

11.2.4 A Primer on Furnaces and Properties

Bundling different residues into one use in Table 11.2 does not contradict our attention to detail in Table 11.1. The properties of individual combustion residues need to be taken into account even when considering the same use. All quantitative information on the properties of ash that has been determined in the Ash Programme as well as all information from Swedish R&D projects on ash is stored in a database, Allaska. It is available in Swedish and English at http://www.askprogrammet.com.[1]

11.2.4.1 Pulverised Fuel Furnaces

Pulverised fuel (PF) furnaces are often very large furnaces converted from furnaces for coal or oil to biomass. The fuel must be ground finely for it to be injected into the

[1]The Web site of the Ash Programme will soon be incorporated into that of Värmeforsk (http://www.varmeforsk.se), but visitors will be redirected to the Värmeforsk Web site.

burners. This is not a very common type, as one usually prefers to retrofit such a furnace with a grate or to convert it to a fluidised bed furnace. However, when there are severe constraints on space in an existing combustion plant, a new PF furnace may be attractive.

The major part of the residues is fly ash. Having passed through high temperatures, 1,200°C and more, fly ash consists of small glassy particles that can yield pozzolanic reactions as in Portland cement. Biomass fly ash is not as good a binding material as coal fly ash, but it is very suitable for road building.

In converted coal or oil furnaces, the bottom ash usually has high contents of unburned carbon and it may be used as fuel in fluidised bed furnaces. It is a poor road building material because of its high water uptake. However, it may be used as low-quality filling material.

11.2.4.2 Grate Furnaces

The grate furnace design is the most common in Sweden up to approximately 100 MW fuel, and the only one for capacities below 20 MW fuel (approximately 6 MW electricity)[2]. The ash has been through high temperatures, but not as high as in PF furnaces. The major part of the residues is bottom ash, and various fly ashes are a minor part.

In a PF furnace, the quantities of fly ash are so much larger than those of bottom ash that the composition of the fly ash corresponds very closely to that of the ashed fuel. With a grate furnace, however, the elements in the ash will be fractionated into several streams of residues. With a grate at more than 1,000°C, volatile elements and their compounds will concentrate in the fly ashes. If heavy metals are an issue, as in the regulations on recycling ash to forest soils, this may pose a problem with fly ash.

The bottom ash is a good road building material, and it is even better if the ash has been burned out and sintered. Even if its water absorption in laboratory tests is quite high, the road will be of good quality and will withstand freeze–thaw cycles. Because of their binding properties, the fly ashes are good building materials for roads. Tests have shown that up to 50% of the cement in stope mine filling may be replaced with these ashes.

11.2.4.3 Fluidised Bed Furnaces

The two types of fluidised bed furnaces commonly distinguished are the circulating fluidised bed furnace and the bubbling fluidised bed furnace. The bed material, usually sand, is finer for the circulating fluidised bed furnace than for the bubbling fluidised bed furnace, 0.3 and 2 mm respectively. Part of the bed is bled out to keep an acceptable chemistry in the bed: alkalis tend to dissolve in the sand and reduce

[2]There are other types with special designs, e.g. cyclone furnaces, but they are not common.

the temperature at which the particles sinter together. The combustion temperature is low, approximately 850°C.

The largest part of the residue is fly ash. The proportion of bottom ash depends on how much material is bled from the bottom of the bed. Both streams of ash contain not only ash from the fuel but also bed material, oversized material in the bottom ash and fine particles from attrition in the fly ash.

The combustion temperature is too low to allow the Portland cement reactions in the fly ash (these require at least 1,400°C). However, they still have binding properties through other reactions which are not fully known. Because of the low combustion temperatures, the particles have irregular shapes. The compressive strength is good even in the green stage if they have been well packed. Among other properties, this makes them very good materials for road building. Mixed in equal proportions with digested sewage sludge, they also provide dense sealing layers to landfills.

The bottom ash contains mostly bed material. It is often too fine and has too narrow a size distribution to be useful in construction, other than as low-quality fill material. However, under certain circumstances, e.g. absence of binding properties, it could be used as backfill material in trenches for piping.

11.3 Recycling Wood Ash to Forest Soils

Sustainability in the use of solid woody biofuels requires that forestry is sustainable. Conventional forestry has an acidifying effect on soils, and intensive harvesting has an even greater effect. "Intensive" is taken to mean whole-tree-harvesting, i.e. timber and pulp wood as well as logging residues. "Sustainability" implies that the mineral nutrients removed in an intensive harvest are returned to the forest soil, e.g. as ash. A good yield of biomass in the future is an important motive, but maybe even more important is to restore the buffering capacity of the soil.

Compensation with ash after harvesting logging residues is an environmental measure encouraged and regulated in Sweden. The details can be found in the recommendations from the Swedish Forest Agency (2002, 2008), which are the fruit of extensive research activities stretching back to the 1980s and financed largely by the government. The environmental impact assessment of removing logging residues published by the National Forest Agency presents the collected scientific basis for the recommendations (Egnell et al. 1998).

To return the desired quantities of mineral nutrients, one needs to keep track of the quantities of logging residues actually removed, which vary depending on soil fertility, essence, etc., and the ash content. Regarding the composition of ash, the basic principle behind the recommendations is that what has been taken away may be returned, nutrients as well as less desirable constituents. A guideline on limit concentrations for elements in ash is then obtained; see Table 11.3.

Table 11.3 The recommended Swedish limit values for nutrients and trace elements in woody biomass ash to be spread to forest soils as a compensation for whole-tree harvest (Swedish Forest Agency 2008)

Macronutrients (g/kg DS)										
Element	Ca	Mg	K	P						
Low limit	125	15	30	7						
Trace elements (mg/kg DS)										
Element	As	B	Cd	Cr	Cu	Hg	Ni	Pb	V	Zn
Low limit	–	–	–	–	–	–	–	–	–	500
High limit	30	800	30	100	400	3	70	300	70	7,000

Table 11.4 Analyses of the concentration of trace elements (mg/kg DS) in the ash of tree parts (H. Eriksson and Swedish Forest Agency, personal communication)

Tree part	Value	As	B	Cd	Cr	Cu	Hg	Ni	Pb	V	Zn
	Limit	30	800	30	100	400	3	70	300	70	7,000
Pine stem	Median	15	342	68	100	313	2	59	24	2	2,928
	85th p	15	540	90	144	437	2	170	133	2	4,305
Spruce stem	Median	12	456	22	29	257	2	39	39	2	3,920
	85th p	32	595	39	110	422	4	84	99	4	5,200
Pine bark	Median	2	337	23	19	127	1	17	21	3	1,707
	85th p	3	422	33	26	163	2	47	37	8	2,448
Spruce bark	Median	2	408	10	13	113	1	37	29	5	4,851
	85th p	3	495	22	22	157	2	71	59	10	5,983
Pine branch	Median	3	615	27	47	276	1	15	81	14	3,615
	85th p	4	774	38	69	382	2	39	139	20	5,397
Spruce branch	Median	5	371	7	42	177	2	20	216	35	5,210
	85th p	5	434	13	65	220	3	76	311	44	6,663

85th p 85th percentile

A consequence of the governing principle of balance is that the tolerable concentration of certain elements in ash is higher than if ash had been considered as a fertiliser. In the latter case, everything in the ash is a net addition to the load on the soil, and the net increase of the concentration of these elements in the soil should be limited. When compensating, one need only to see that the net addition is negligible. Table 11.4 (H. Eriksson and Swedish Forest Agency, personal communication) provides a background for Table 11.3: it summarises the results of analyses of ashed softwood with respect to these contentious trace elements. For each element, the median of the available set of data and the 85th percentile are presented; the number of data ranges from a dozen to slightly more than 100.

The main result of studies of the environmental impact of ash is that these high concentrations of trace elements are not an important issue with the low doses of ash that are being spread for compensation purposes (Egnell et al. 1998). The basicity is a greater problem: ash may not be used fresh from the furnace. Ash must be stabilised by mixing it with water and allowed to mature during storage so that its pH is lowered. To these recommendations, one should add the regulation from the Radiation Protection Authorities that limit the concentration of caesium-137 in ash that is to be used. This is of course owing to the impact from the Chernobyl accident.

Table 11.5 Minimum values in the Swedish recommendations for ashes recycled to forests and median values of concentrations in Allaska (g/kg DS) for main elements in clean wood ash

Element	Minimum value	Median value wood fly ash, all furnaces	Median value wood bottom ash, grate furnace
Ca	125	172	131
Mg	15	18	16
K	30	42	38
P	7	8	7
Zn	0.5	3	0.33

Table 11.6 Maximum values of concentrations in the Swedish recommendations for ashes recycled to forests (mg/kg DS) and median values in Allaska for trace elements in clean wood ash

Trace element	Maximum value	Fly ash, grate furnace	Bottom ash, grate furnace	Fly ash, fluidised bed furnace	Bottom ash, fluidised bed furnace
As	30	9.8	7.6	14	36.5
B	800	230	205	295	290
Cd	30	10	1	9	0.3
Cr	100	147	265	77	60.5
Cu	400	120	120	145	135
Hg	3	0.2	0.01	0.4	0.012
Ni	70	21	12	47	18
Pb	300	145	89	122	74
V	70	17	43	36	30
Zn	7,000	2,100	461	1,460	151

In principle, it should always be possible to spread ash from clean wood to forest soils. However, it is unavoidable that some residues will have concentrations in excess of these recommended limits. Some trees (birch, pine and willow) may naturally contain excessively high quantities of cadmium. Fractionation in the furnace will remove cadmium and caesium from the bottom ash of a grate furnace and concentrate them in the fly ash. Fractionation does though also reduce the concentration of potassium and zinc in the bottom ash. Other elements that may cause concern are chromium and sometimes nickel, a result of corrosion in the furnaces or in the fuel handling equipment. An illustration of the span of actual concentrations is given in Table 11.5 for some nutrients and in Table 11.6 for trace elements.

The only combustion residues that consistently meet the requirements are fly ashes from FB furnaces. One should remember that their content of objectionable substances is diluted by the bed material that is entrained with the fly ash.

Phosphorus is especially important as conventional harvest already creates a deficiency in many stands, and the deficiency is accentuated by intensive harvesting where logging residues are removed. The experience acquired hitherto is that the phosphorus content of the combustion residues generally does not reach the desired levels. Zinc is also problematic, as the zinc content of bottom ashes seldom exceeds the lowest level in the recommendations.

A guideline for the classification of ash from solid biofuels and peat has been published by Nordtest (Haglund and Expert Group 2008) targeting its use in recycling and fertilising. A table comparing guideline values for elements in ash in the Nordic countries is provided there. If an ash does not meet all of the requirements for spreading on soils as a compensation measure, it should be used as material in civil works.

However well accepted by authorities, compensation has not yet really taken off: only part of the logging sites that should receive a compensatory dose of ash has actually received one. The Swedish Forest Agency has intensified the distribution of information to stakeholders in order to push adoption of compensation with ash, for example through the EU-Life project RecAsh (Emilsson 2006). The Web site of RecAsh has been closed down, but all information is now available in Swedish, Finnish and English on the Web site of the Swedish Forest Agency (http://www.skogsstyrelsen.se).

The Ash Programme has chosen to complement the RecAsh project as well as the extensive R&D programmes on wood ash by focusing on biomass growth. It was perceived that one essential non-technical barrier is the absence of an economic incentive: spreading ash to soils in doses of a couple of tonnes per hectare is costly but the forest owner is not to expect an immediate return as increased growth. If increased growth could nevertheless be demonstrated for at least some circumstances, one could expect an increased general interest. However, this is not an easy proposition as growth in Swedish forests in most cases is limited by the availability of nitrogen.

In these discussions one has to consider mineral soils separately from organic soils (peat lands), as they are two different cases. Much of the R&D leading to the official recommendations dealt with mineral soils. Although some increased growth was observed on fertile sites, some reduced growth could be observed on less fertile sites. None of the figures in isolated studies are statistically significant, but the general trend is convincing. However, public opinion has mostly retained the negative part of the conclusions, i.e. decreased growth. On the other hand, there is plenty of historical evidence for increased growth of drained peat lands as a result of applying ash. The adverse effects of ash on mineral soils are now well known, but it is feared that applying ash to peat lands will, for example, increase the emission of greenhouse gases.

There is controversy regarding the causes for increased growth on mineral soil. In one school of thought, the basicity of ash increases the availability of nitrogen to vegetation. In the other school of thought, increased growth is a result of an improved nutrient supply, primarily phosphorus. Older experimental fields have been revisited within the Ash Programme and new ones have been established (Sikström et al. 2006, 2009a; Thelin 2006, 2009). Evidence is coming in, but slowly. So far, none of the interpretations have been invalidated.

Peat land is usually taken to be only the present peat bogs or peat cutovers. However, in the 1950s and 1960s almost one million hectares was drained and afforested in Sweden. Growth is very variable as peat lands are almost always deficient in several mineral nutrients. In a prestudy by Hånell and Magnusson

(2005), it was found that spreading ash as a fertiliser on approximately 190,000 ha, i.e. compensating for the future harvest before it is actually carried out, could yield a short-term profit. Larger areas could benefit from an application of ash, but it will take more time to obtain harvestable trees.

The priority of the Ash Programme is to address possible adverse effects. In a series of studies, the effect of ash on water chemistry, on the production of greenhouse gases (methane, nitrous oxide, carbon dioxide), on biodiversity and on microbial mass was investigated (Sikström et al. 2006, 2009b), as was also done by other authors (Kuba et al. 2008; Bougnom et al. 2010). The results do not indicate any important negative effect: the production of methane is mostly unaffected, that of carbon dioxide decreases, biodiversity increases...

The Ash Programme is now shifting its emphasis towards growth on fertile mineral soil. Peat lands remain interesting targets in northern Sweden, but nature conservation requirements as well as difficulties to use modern and heavy machines on such soils make them less interesting in southern Sweden.

11.4 Civil Works and Landfills

11.4.1 Landfills and Mining Waste Deposits

Combustion residues are used to provide a barrier against the penetration of water or oxygen, or both, into the body of the landfill or of the heap or impoundment of mining waste. The properties that make fly ash from solid biofuels suitable are their high pH and their self-binding properties. The potential for use in several contexts is quite high and the demand could easily exceed the availability of suitable residues.

Fly ash mixed into sewage sludge in equal dry substance proportions raises the pH of the sludge, thus preventing its biological degradation. The percolation rates achieved in field experiments with fly ash mixed into sewage sludge are on the order of 12 $l/m^2/year$, both initially and after a few years, which is sufficiently low for sealing layers on landfills for non-hazardous waste (Carling et al. 2006; Mácsik et al. 2005). Its shear strength is acceptable and it withstands settling in the body of the landfill.

On the Tveta landfill, the functional requirements on a sealing layer are fulfilled using monolithic layers of ash through diffusional processes, i.e. particle size distribution, moisture content and reactivity of the combustion residues, all contributing to the minimisation of the pore volume (Tham and Andreas 2008).

A sealing layer of fly ash from solid biofuels, covered by a protective layer of vegetated sewage sludge, prevents oxygen from reaching sulphidic tailings from mining. The experimental object is the 80-ha large tailings impoundment of Gillervattnet, where an experimental field covering 3 ha was established in 2003–2005. The hardened layer of ash and the toxicity of the fresh ash prevent roots from breaking through the sealing layer and providing channels for air entry.

However, reed canary grass has some capacity to weaken even hardened fly ash sealing layers with a resistance of approximately 5 MPa. The study suggests that the secretion of saccharides by some plant roots may contribute to this effect (Greger et al. 2006, 2009). It is thus best to avoid these particular plants.

11.4.2 Utilisation in Civil Works

Although roads are not a particularly suitable use for residues for logistical reasons, they are the type of civil works with the best developed set of requirements, against which constructions with combustion residues may be assessed. Fly ash, preferably from solid biofuels or sludges from the pulp and paper industry, has been used in non-surfaced gravel roads and bottom ash is now also beginning to be used.

Non-surfaced roads have been built using fly ash in Sweden for some time. However, through the introduction of the Finnish experience with this technology (Lahtinen 2001) as the starting point within the Ash Programme, a convenient impulse was given to renewed development. In successive projects, fly ash from biomass has been characterised, recipes have been developed and a few test roads have been built (Mácsik and Svedberg 2006; Mácsik et al. 2009). A short summary of the results is as follows: bearing capacity and freeze–thaw resistance have increased, fly ash replaces natural materials of about twice its volume, which leads to a significant reduction in weight and height. For the good properties of biomass fly ash to be optimally exploited, and for conservation of fly ash resources, ash should be used to stabilise bad or worn-out road materials. Adverse impacts on the environment could not be observed during the monitoring of construction and use of the roads.

Bottom ash from solid biofuels is also used for roadwork. An example is a private road north of Norrtälje, where approximately 5,000 t of mixed bottom and fly ash from a grate furnace have been used since 2006; see Fig. 11.1. It was possible to run 40-t lorries on the road even when it was flooded by melting snow, which is remarkably good. Infiltration in the body of the work is very slow, which indicates that the environmental impact should be minor. The impact of an ash pile stored for 7 years on the place has been investigated: the uptake of heavy metals in ash by plants and berries did not lead to any increased levels in the plants.

The ash of solid biofuels also has binding properties and it has been utilised in concrete applications. One project involved replacing Portland cement in panel stope mine filling, where large blocks or "stopes" of ore are removed, creating a large cavity. These stopes are backfilled using concrete (cement and mine tailings) to stabilise the mine. In full-scale trials, biomass fly ash from grate furnaces could replace 50% of the Portland cement. The other use of solid biofuel ashes demonstrated in the Ash Programme is as filler in low-quality concrete. However, the chloride content of the ash may pose corrosion problems for steel reinforcement bars.

Fig. 11.1 Photograph of a stretch of the gravel road built at Tranbol using mixed wood ashes

11.5 Environmental Permitting for Civil Works

The key to utilising combustion residues as well as other mineral wastes in civil works and keeping them out of landfills is the assessment of the environmental impact in the permitting procedures. This was one of the main reasons for creating the Ash Programme: a method of computing impact had to be developed and numerical values for the properties have to be fed into the method. This work was performed in conjunction with environmental authorities, but this does not mean that these authorities reach the same conclusion in their assessment.

Initially, concepts similar to the "end of waste" presently discussed in the EU were considered, but as an assessment still has to be done, the method of the Swedish Environmental Protection Agency for assessing the impact of contaminated soil on health and the environment was chosen as a starting point and adapted to the pilot cases (Bendz et al. 2006):

- A non-surfaced road in a forest, with a comparatively thin layer of ash
- A surfaced road with MSWI bottom ash in the subbase, with a comparatively thick ash layer

The purpose of the assessment is to define the boundary between a low risk level and a not low risk level (in legalese terms, as "no risk" does not exist and "high risk" will not be allowed). Below the boundary, a simplified procedure could be defended, for example only giving notice to the environmental authorities. Above the boundary, a full permitting procedure would of course be necessary, with a detailed analysis of the expected local environmental impact.

All mechanisms for dispersal from the body of the road and for human exposure were described in the model. The model is conservative: a plausible worst-case scenario is assumed, yielding rather large safety margins. For example, the most exposed person is assumed to live all his or her life within 20 m from the road, 30% of this person's intake of vegetables is home-grown close to the road (hardly washing them) and when the road is disused after some 60 years, it is used as a recreation area by adults and children, assuming 40 windy days per year and person.

The result of these first computations is that dispersal of dust yields the dominating health risk (Bendz et al. 2006). Criteria based only on leaching to groundwater yield significantly larger limit values for, for example, heavy metals in the combustion residues. In the simulations, even with MSWI bottom ash, leaching from several roads yields insignificant increases in heavy metal content in the recipients of two catchment areas (Wik 2009).

To put this result in perspective, the composition and leaching properties of most combustion residues are such that, if correctly used, these residues present a "low risk" to health and the environment. The only ones that cannot satisfy the upper limit values for low risk are APC residues from MSWI and fly ash or APC residues from combustion of impregnated wood.

The low risk limit values for some of the trace elements are shown in Table 11.7. Two references are given in Table 11.7 for the sake of comparison. The first reference is a set of limit values derived by the Swedish Environmental Protection Agency after the Ash Programme had published its proposal (Swedish Environmental Protection Agency 2010). The purpose of this set is to define a boundary below which the user of a waste material does not even need to give notice of his or her use of these materials. These values are substantially lower as they represent the 90th percentile of concentrations in soil. The boundary defined by the Swedish Environmental Protection Agency and that proposed by Bendz et al. (2006) and later updated (Bendz et al. 2009) are not the same. The second reference is the recommended limit values for ash spread to forest soils as nutrient compensation (Swedish Forest Agency 2008).

Note the comparatively low concentration found for arsenic, 15 mg/kg dry substances, the immediate reason for which is that arsenic is genotoxic. One should keep in mind though that the results in Table 11.7 are the first results based on conservative models, conservative assumptions and uncertain data. For the time being, it is recommended that ash with an arsenic content in excess of this value should not be left on the surface when the road is abandoned.

Table 11.7 Maximum tolerable concentration (mg/kg DS) of key trace elements in ash for the boundary between low risk and not low risk for health and the environment, values computed by Bendz et al. (2009) for uses of combustion residues in civil works

	Pb	Cd	Hg	Zn	As
Low local risk in gravel roads	1,400	330	80	>25,000	55
Low risk when ash is exposed at the surface of a disused road	1,500	60	60	>50,000	15
Free use according to the Swedish Environmental Protection Authority	20	0.2	0.1	120	10
Swedish Forest Agency, compensation for whole-tree harvesting	300	30	3	7,000	30

11.6 Ecotoxicity

Ecotoxicity is a key property from a regulatory point of view, H14 in the European Waste Catalogue. Although the EU is expected to issue rules on criteria 2011, today there is no legally binding rule. There is no accepted test procedure for ecotoxicity yet, and an H14 value is computed using knowledge of the chemical composition of the waste and of the ecotoxicity of the pure chemical substances.

There is a difference of opinion in Sweden on the level at which a waste is to be considered ecotoxic. When a chemical product contains more than 2.5% of an ecotoxic substance, it is hazardous enough for organisms to warrant labelling as such. Analyses of ash usually yield the elemental composition, very seldom the concentration of chemical compounds. A study for the Ash Programme suggested a conservative selection of substances for the key trace elements and recommended that this value, 2.5%, be used to determine whether an ash is ecotoxic or not. The environmental authorities are of the opinion that the limit should be 0.25% because this is the limit for aquatic organisms.

The key element in this discussion is zinc, an essential nutrient. To be conservative and not underestimate the ecotoxicity, the above-mentioned study assumed zinc was present as zinc oxide. At approximately the same time, zinc oxide was reclassified as ecotoxic, which has severe consequences for bioash. The zinc content in clean wood ash recommended by the Swedish Forest Agency is up to 7,000 mg/kg, i.e. 0.7%, which is larger than 0.25%. Concentrations of approximately 0.4% are not uncommon for bark ash.

One way of tackling this difficulty is to examine whether zinc oxide is a reasonable choice as a reference substance after all, rather than just a conservative one. Three investigations should be mentioned here:

- Van der Sloot at ECN utilised the database LeachXS on leaching properties of MSWI residues and concluded that zinc is present mostly as willemite, a zinc silicate (H. van der Sloot, oral communication).
- Sjöblom (2007) reexamined the chemical thermodynamics of zinc in combustion residues, in particular the solubility equilibria, and concluded that zinc is present probably as franklinite, a mixed iron and zinc oxide, rather than willemite.
- Steenari and Norén (2008) studied the binding distances for zinc in different combustion residues using extended X-ray absorption fine structure spectroscopy, and found that the distances agree with zinc being bound in silicates or aluminates. They also found that in fresh ash, some of the more soluble zinc compounds are also present in minor quantities. When ash is wetted, zinc is rapidly redistributed to insoluble compounds.

Thus, zinc does not occur as oxide in combustion residues, but rather in much more stable compounds. Choosing oxide in the ecotoxicity assessment was overly conservative and significantly overestimates the environmental impact of zinc in ash. Settling for a realistic model compound is not straightforward: it seems more probable that zinc is found as willemite or a silicate, but there is no official assessment of these compounds. On the other hand, less probable franklinite but with similar properties has been assessed as non-ecotoxic.

At the end of the day, when there are doubts, an ecotoxicity test will clinch the discussion. In a study co-financed by the Ash Programme, the Swedish Waste Management Association and the Swedish Environmental Protection Agency, a battery of tests were performed on different combustion residues (Stiernström et al. 2009). Care was taken to choose organisms that naturally occur in brackish waters rather than the freshwater organisms that are sensitive to the salt concentration.

At a liquid-to-solid leaching ratio of 10 (waste test condition to be on the conservative side, but for chemical products the liquid-to-solid leaching ratio should be 10,000 according to OECD good practice) almost all residues were ecotoxic. A more detailed examination revealed that the ecotoxicity is due to the high concentrations of potassium and calcium (nutrients) and of aluminium. This is scarcely a reason to consider the residues as ecotoxic.

Results from tests of bottom ash from MSWI are illustrative. Ash which had been stored and allowed to mature for 3 months was somewhat ecotoxic. Ash which had been aged and exposed to the weather (actual liquid-to-solid leaching ratio of 3.4, rainwater as leachant) for 12 years was not ecotoxic at all to the three organisms tested. And yet, its composition had not changed, except for a lower content of chlorides and probably of sulphates, too.

11.7 Conclusion

Projects within the Ash Programme have contributed increased knowledge of the properties of combustion residues as well as demonstrated large-scale uses for residues. The Ash Programme has been active since 2002, and it is quite probable that it will be extended for another 3 years at the end of 2011.

With ash from clean solid biofuels, priority should be given to recycling the ash to forest soils as a compensation for the extraction of mineral nutrients in an intensive harvesting, i.e. whole-tree harvesting. If that is not possible, the residues should be used in civil works, e.g. for building roads or hard surfaces. One should put all such ash to use, instead of rejecting particular streams.

The Ash Programme has striven for a common understanding of the level of risk for human health and the environment when using ash in civil works. Boundaries between low risk and not low risk have been suggested. The Swedish Environmental Protection Agency has developed another boundary, below which one does not even need to pay notice to the environmental authorities.

All reports are public and are available from Värmeforsk's Web site (http://www.varmeforsk.se). The language is Swedish, but there is always a summary in English.

References

Bendz D, Wik O, Elert M, Håkansson K (2006) Environmental guidelines for reuse of ash in civil engineering applications – phase 2 (in Swedish). Report no. 979. Swedish Thermal Engineering Research Institute, Stockholm. Available via http://www.varmeforsk.se

Bendz D, Wik O, Jones C, Pettersson M, Elert M (2009) Environmental guidelines for reuse of ash in civil engineering applications – including criteria for Sb and As (in Swedish). Report no. 1110. Swedish Thermal Engineering Research Institute, Stockholm.. Available via http://www.varmeforsk.se

Bjurström H, Herbert R (2009) The Swedish Ash Programme 2002–2008, Biomass, wastes, peat – any solid fuel but coal. Report no. 1111. Swedish Thermal Engineering Research Institute, Stockholm.. Available via http://www.varmeforsk.se

Bjurström H, Herbert R, Ribbing C, Tham G (2009) Using ash in a sustainable society – Swedish R&D programme 2002–2008. In: Proceedings of Sardinia 2009, 12th International Waste Management and Landfill Symposium, Cagliari

Bougnom BP, Knapp BA, Etoa FX, Elhottová D, Němcová A, Insam H (2010) Designer composts for ameliorating acid tropical soils: effects on the soil microbiota. Appl Soil Ecol 45:319–324

Carling M, Ländell M, Håkansson K, Myrhede E (2006) Covering of landfills with ash and sludge – experience from field tests (in Swedish). Report no. 948. Swedish Thermal Engineering Research Institute, Stockholm. Available via http://www.varmeforsk.se

Egnell G, Nohrstedts H-Ö, Weslien J, Westling O, Örlander G (1998) Environmental impact assessment of harvesting logging residues, spreading ash and other nutrient compensation (in Swedish). Report no. 1 1988. Swedish Forest Agency, Jönköping. Available via http://www.skogsstyrelsen.se

Emilsson S (2006) International handbook – from extraction of forest fuels to ash recycling. Swedish Forest Agency, Jönköping. Available via http://www.skogsstyrelsen.se

Greger M, Neuschütz C, Isaksson K-E (2006) Sealing layer of fly ash and sewage sludge and vegetation establishment in treatment of mine tailings impoundment (in Swedish). Report no. 959. Swedish Thermal Engineering Research Institute, Stockholm. Available via http://www.varmeforsk.se

Greger M, Neuschütz C, Isaksson K-E (2009) Influence of vegetation and sewage sludge on sealing layers of fly ash in post-treatment of mine tailings impoundment (in Swedish). Report no. 1098. Swedish Thermal Engineering Research Institute, Stockholm. Available via http://www.varmeforsk.se

Haglund N, Expert Group (2008) Guideline for classification of ash from solid biofuels and peat utilised for recycling and fertilizing in forestry and agriculture. Nordtest technical report no. 613. Nordic Innovation Centre, Oslo. Available via http://www.nordicinnovation.net

Hånell B, Magnusson T (2005) An evaluation of land suitability for forest fertilization with biofuel ash on organic soils in Sweden. For Ecol Manage 209:43–55

Kuba T, Tschöll A, Partl C, Meyer K, Insam H (2008) Wood ash admixture to organic wastes improves compost and its performance. Agric Ecosyst Environ 127:43–49

Lahtinen P (2001) Fly ash mixtures as flexible structural materials for low-volume roads. Report no. 70/2001. Finnish Road Administration, Helsinki. http://alk.tiehallinto.fi/kirjasto/ejulk_2.htm#2001

Mácsik J, Svedberg B (2006) Gravel road stabilization of Ehnsjövägen Hallstavik (in Swedish). Report no 968. Swedish Thermal Engineering Research Institute, Stockholm. Available via www.varmeforsk.se

Mácsik J, Maurice C, Mossakowska A, Eklund C, Erlandsson Å (2005) Pilot study with fly ash stabilised sewage sludge as hydraulic barrier layer (in Swedish). Report no. 942. Swedish Thermal Engineering Research Institute, Stockholm. Available via http://www.varmeforsk.se

Mácsik J, Erlandsson Å, Wexell B-Å (2009) Fly ash and green liquor as binder in gravel road stabilization – pilot study at Iggesund (in Swedish). Report no. 1101. Swedish Thermal Engineering Research Institute, Stockholm. Available via http://www.varmeforsk.se

Sikström U, Ernfors M, Jacobsson S, Klemedtsson L, Nilsson M, Ring E (2006) Addition of ash on drained forested peatlands in southern Sweden – effects on forest production, fluxes of greenhouse gases and water chemistry (in Swedish). Report no. 974. Swedish Thermal Engineering Research Institute, Stockholm. Available via http://www.varmeforsk.se

Sikström U, Jacobsson S, Johansson U, Kokkola M, Saarsalmi A, Holt-Hansen K (2009a) Long-term growth responses of ash addition and liming – preliminary results from a pilot study (in Swedish). Report no. 1107. Swedish Thermal Engineering Research Institute, Stockholm. Available via http://www.varmeforsk.se

Sikström U, Björk RB, Ring E, Ernfors M, Jacobsson S, Nilsson M, Klemedtsson L (2009b) Addition of ash on drained forested peatlands in southern Sweden – effects on forest production, fluxes of greenhouse gases, peat properties, understorey vegetation and groundwater chemistry (in Swedish). Report no. 1109. Swedish Thermal Engineering Research Institute, Stockholm. Available via http://www.varmeforsk.se

Sjöblom R (2007) Basis for choosing a reference substance for zinc when classifying a waste according the waste ordinance (in Swedish). Report no. F2007:01. Swedish Waste Management Association, Malmö. Available via http://www.avfallsverige.se

Steenari B-M, Norén K (2008) Speciation of zinc in ash investigated by X-ray absorption spectroscopy (in Swedish). Report no. 1063. Swedish Thermal Engineering Research Institute, Stockholm. Available via http://www.varmeforsk.se

Stiernström S, Hemström K, Wik O, Carlsson G, Breitholz M (2009) Methodology for classification of the H14 criterion according to the directive 2008/98/EC on waste (in Swedish). Report no. 1092. Swedish Thermal Engineering Research Institute, Stockholm. Available via http://www.varmeforsk.se

Swedish Environmental Protection Agency (2010) Reusing waste in civil works (in Swedish). Handbook 2010:1. Swedish Environmental Protection Agency, Stockholm. Available via http://www.naturvardsverket.se

Swedish Forest Agency (2002) Recommendations for the extraction of forest fuel and compensation fertilizing. Communication no. 3·2002. Swedish Forest Agency, Jönköping. Available via http://www.skogsstyrelsen.se

Swedish Forest Agency (2008) Recommendations for the extraction of logging residues and ash recycling (in Swedish). Communication no. 2·2008. Swedish Forest Agency, Jönköping. Available via http://www.skogsstyrelsen.se

Tham G, Andreas L (2008) Results from full-scale application of ashes and other residuals in the final cover of the Tveta landfill (in Swedish). Report no. 1064. Swedish Thermal Engineering Research Institute, Stockholm. Available via http://www.varmeforsk.se

Thelin G (2006) Recycling ash to beech and spruce stands – effects on nutrition, growth, nitrogen dynamics and carbon balance (in Swedish). Report no. 965. Swedish Thermal Engineering Research Institute, Stockholm. Available via http://www.varmeforsk.se

Thelin G (2009) Bio energy production in birch and hybrid aspen after addition of residue based fertilizers – establishment of fertilization trials (in Swedish). Report no. 1094. Swedish Thermal Engineering Research Institute, Stockholm. Available via http://www.varmeforsk.se

Wik O (2009) Regional risk assessment of reuse of ash in civil engineering applications (in Swedish). Report no. 1113. Swedish Thermal Engineering Research Institute, Stockholm. Available via http://www.varmeforsk.se